Ecological Studies, Vol. 121

Analysis and Synthesis

Edited by

G. Heldmaier, Marburg, FRG
O.L. Lange, Würzburg, FRG
H.A. Mooney, Stanford, USA
U. Sommer, Kiel, FRG

Ecological Studies

Volumes published since 1990 are listed at the end of this book.

Springer

Berlin
Heidelberg
New York
Barcelona
Budapest
Hong Kong
London
Milan
Paris
Santa Clara
Singapore
Tokyo

O.T. Solbrig E. Medina J.F. Silva (Eds.)

Biodiversity and Savanna Ecosystem Processes

A Global Perspective

With 40 Figures

Springer

Prof. Dr. Otto T. Solbrig
Harvard University
Department of Organismic
and Evolutionary Biology
22 Divinity Avenue
Cambridge, Massachusetts, 02138
USA

Prof. Dr. Ernesto Medina
Centro de Ecologia
Instituto Venezelano
de Investigaciones Cientificas
Aptdo. 21827
1020-A Caracas
Venezuela

Prof.Dr. Juan F. Silva
Universidad de los Andes
CIELAT, Facultad de Ciencias
Merida
Venezuela

ISBN-13: 978-3-642-78971-7

Library of Congress Catolging-in-Publication Data
Biodiversity and savanna ecosystem processes : a global perspective /
O.T. Solbrig, E. Medina, J.F. Silva (eds.).
p. cm.– (Ecological studies : v. 121)
Includes bibliographical references and index.
ISBN-13: 978-3-642-78971-7 e-ISBN-13: 978-3-642-78969-4
DOI: 10.1007/978-3-642-78969-4
1. Savanna ecology. 2. Biodiversity. 3. Species diversity.
I. Solbrig, Otto Thomas. II. Medina, Ernesto. III. Silva, J.F.
(JUan F.), 1941- . IV Series.
QH541.5.P7B56 1995
574.5'2652–dc20 95-34654

Softcover repint of the hardcover 1st edition 1996

Cover design: D & P, Heidelberg
Typesetting: Dr. Kurt Darms, Bevern
SPIN 10466664 31/3137-5 4 3 2 1 0 - Printed on acid-free paper

Preface

Ecologists and population biologists have been concerned with the effects of the physical and biological characteristics of the ecosystem on species behavior and diversity. Yet as in any system, the reverse is also true: species and their behavior will also affect the physical and biological characteristics of the ecosystem. This volume addresses this latter question in the context of tropical savanna ecosystems.

This work is a joint contribution of the Responses of Savannas to Stress and Disturbance (RSSD) Program of the Decade of the Tropics, a joint IUBS-UNESCO International Program, and of the DIVERSITAS Program, a joint IUBS-SCOPE-UNESCO International Program. The former, as its name implies, was primarily concerned with savanna function in the face of natural and human-made disturbances; the second addresses the effects that changes in biodiversity have on ecosystem function.

The chapters presented in this volume resulted from a conference held in Brasilia, Brazil on „The Role of Biodiversity in the Function of Savanna Ecosystems“, from the 24th and the 28th of May, 1993, sponsored by the International Union of Biological Sciences (IUBS), the Scientific Committee for Problems of the Environment (SCOPE); the United Nations Educational, Scientific and Cultural Organization (UNESCO), and the Instituto Brasileiro do Meio Ambiente e dos Recursos Naturais Renovaveis (IBAMA). We wish to thank the sponsors heartily, as well as the National Science Foundation of the United States, The Fundación Polar of Venezuela, and The Consejo Nacional de Ciencia y Tecnologia (CONICIT) of Venezuela for additional travel support.

The book is divided into two sections. The first which we have entitled „Biodiversity and Savanna Structure and Function“ consists of nine chapters that were solicited by the editors prior to the conference. The second part, called „Summary and Areas for Future Research“ was written after the conference and deals with the issues as the participants in the Brasilia conference saw them. We conclude with a chapter that summarizes the major conclusions regarding the function of Biodiversity in the function of Savanna Ecosystems.

We wish to thank all the participants, but above all Dr. Braulio Dias and Dr. Adriana Moreira, formerly with IBAMA, our gracious hosts who organized a very fruitful and rewarding meeting in very trying circumstances. We also wish to thank the authors for their patience and Ms. Patsy Phillips for her help in preparing the manuscript.

November 1995

Otto T. Solbrig
Ernesto Medina
Juan F. Silva

Contents

Biodiversity and Savanna Structure and Function

Summary and Areas for Future Research

Contributors

STEVE ARCHER
Dept. of Rangeland Ecology and Management, Room 225 Animal Industries Building, Texas A&M University, College Station, Texas 77843-2126, USA.

ZDRAVKO BARUCH
Depto. Estudios Ambientales, Universidad Simón Bolívar, Apartado 89000, Caracas 1080, Venezuela

A. JOY BELSKY
1730 S. W. Harbor Way, Apt. 603, Portland, Oregon 97201, USA.

BIBIANA BILBAO
Centro de Ecología, Instituto Venezolano de Investtigaciones Científicas, Aptdo. 21827, Caracas 1020-A, Venezuela

RICHARD W. BRAITHWAITE
CSIRO Division of Wildlife and Ecology, PMB 44, Winnellie Northern Territory, 0821, Australia

LUIS BULLA
Instituto de Zoología Tropical, Facultad de Ciencias, Universidad Central de Venezuela, Apartado 47058, Caracas 1041-A, Venezuela

MIKE COUGHENOUR
Natural Resource Ecology Laboratory, Colorado State University, Fort Collins, Colorado 80523, USA.

CHRISTIANE DALL AGLIO
Depart. Ecology, University of Brasilia, Brasilia, CP 153081, 70910 Brasilia, Brazil

G. WILSON FERNANDEZ
Dept. Biologia Geral, CP 2486, Univ. Federal de Minas Gerais, 30161 Belo Horizonte, MG, Brazil

C. AUGUSTO FRANCO
Dept. Botany, University of Brasilia, 70910 Brasilia, Brazil

IRENE GARAY
Dept. Botanica, Universidade Federal Rural do Rio de Janeiro, Rio de Janeiro, Brazil

MUNDAYATAN HARIDASAN
Dept. of Ecology, University of Brasilia, CP 153081, 70910 Brasilia, Brazil

JOHN HAY
Dept. of Ecology, University of Brasilia, CP 153081, 70910 Brasilia, Brazil

William Hoffmann
Dept. O.E. Biology, Harvard University, 22 Divinity Ave., Cambridge, Massachusetts 02138, USA.

Carlos Klink
SQS 303-I-207, Brasilia, D.F., 70336-000, Brazil

Patrick Lavelle
ORSTOM - Laboratoire d'Ecologie des Sols Tropicaux, 72 Route d'Aulnay 93143 Bondy Cedex, France

Thomas M. Lewinsohn
Laboratório de Interações Insetos-Plantas, IB, Unicamp, 13083-970 Campinas, SP, Brazil and NERC Centre for Population Biology, Ascot, Berkshire, UK

Ernesto Medina
Centro de Ecología, Instituto Venezolano de Investigaciones Científicas, Aptdo. 21827, Caracas 1020-A, Venezuela

Adriana Moreira
Environmental Advisor, USAID Embaixada Americana Av. das Nacoes - Q. 801 - Lote 3 CEP 70.403.900, Brasilia, Brazil

Paulo E. Oliviera
Dept. de Biociências, Univ. Federal de Uberlândia, 38400, Uberlândia, MG, Brazil

Peter W. Price
Department of Biological Sciences, Northern Arizona University, Flagstaff, Arizona 86011-5640, USA

José Felipe Ribeiro
Centro de Pesquisas do Cerrado, (CPAC) EMBRAPA, Planaltina, DF, Brazil

Guillermo Sarmiento
Centro de Investigaciones Ecológicas de los Andes Tropicales (CIELAT), Facultad de Ciencias, ULA, Mérida, Venezuela

Juan F. Silva
Centro de Investigaciones Ecológicas de los Andes Tropicales (CIELAT), Facultad de Ciencias, ULA, Mérida, Venezuela

Otto T. Solbrig
Dept. O.E. Biology, Harvard University, 22 Divinity Ave., Cambridge, Massachusettes, USA

Philip Stott
Dept. of Geography, School of Oriental and African Studies, University of London, Thornhaugh Street, Russell Square, London WCIH OXG

Biodiversity and Savanna Structure and Function

1 The Diversity of the Savanna Ecosystem

1 The Diversity of the Savanna Ecosystem

Otto T. Solbrig

1.1 Introduction

Savannas are defined by Bourlière and Hadley (1983) as "tropical grasslands with scattered tress." Ecosystems with these characteristics are found throughout the tropics in all continents but particularly in the Americas, Africa, and Australia. However, this physiognomic definition encompasses a great deal of diversity in both biotic and abiotic characteristics, in floristic composition and in vegetation history. This chapter presents a summary of this diversity.

Savanna diversity is present at all scales. At an inter-continental scale there are differences in vegetation structure and floristic composition, but across continents there is a similar range of soil types and climatic regimes; within continents regional differences in soils and climate determine the principal types of savannas, while at a local scale, differences in topography, and geomorphology are the principal determinants of local vegetation structure and floristic composition (Solbrig 1993). This hierarchy of effects must be kept in mind as we review the principal savanna regions of the tropics.

1.2 Types and Geographical Extent of Savannas

Savannas, that is tropical ecosystems with a continuous layer of grasses with or without a discontinuous layer of shrubs and/or trees occupy approximately 40% of the surface of the tropics, some 23 million km^2 (Cole 1986).

Ecological Studies, Vol. 121
Solbrig, Medina and Silva (eds) Biodiversity and Savanna Ecosystem Processes

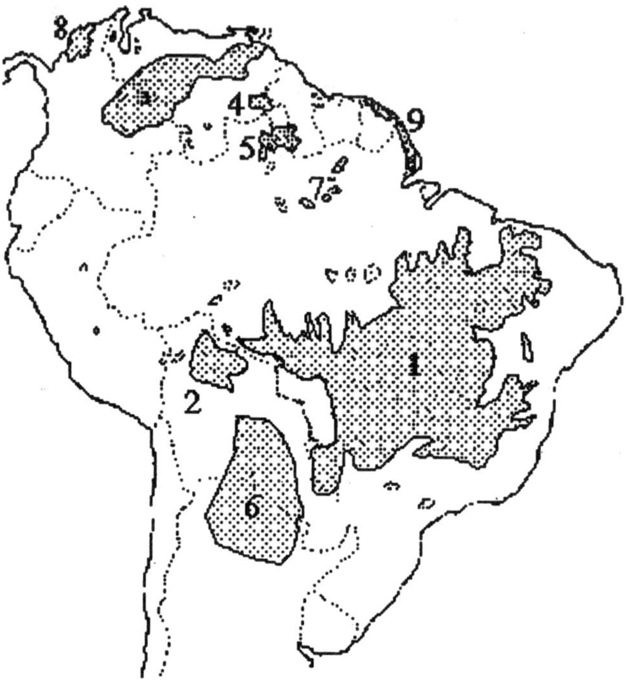

Fig. 1.1. Principal South American Savanna regions. *1* Cerrado; *2* Llanos de Moxos; *3* Llanos del Orinoco; *4* Gran Sabana; *5* Savannas of the Rio Branco-Rupununi; *6* Chaco; *7* Amazonian campos; *8* Llanos of the Magdalena; *9* coastal savannas of the Guayanas. (After Sarmiento 1984)

In South America savannas occur in two large patches north and south of the equator. The principal savanna region south of the equator occurs entirely within Brazil and is locally known as the *cerrado* (Fig. 1.1, Novaes Pinto 1994) encompassing an area of approximately 1.8 million km^2. Physiognomically, the cerrado is formed by a mixture of plant formations that grade into each other and which are known by their local Brazilian names. *Campo limpo* is a pure grassland formation, that turns into *campo sujo* when the grassland is dotted with shrubs and small trees. When the density, and to some extent the size, of the woody vegetation increases, it is first called a *campo cerrado* and then a *cerrado tipico* or just *cerrado*. At this point the savanna is dominated by the woody element, so that it is best described as a low open woodland. This is the most extensive formation of the Brazilian savanna and gives it its name. A last stage in the gradient from open to close formations is the so-called *cerradao*, small dry-forest islands, where grasses, although present, form a minor vegetation element, and which, together with some typical species of the more open formation, also contain a significant number of species that are restricted to the *cerradao*. Cerrado formations cover approximately 85% of the cerrado phytogeographical province. The remainder is occupied by wet forests along streams of water (gallery forests) and other areas with permanent water; it can also be covered by characteristic palm forests of *buriti* and *buritirana* (*Mauritia vinifera* and *M. martiana*). Other special types of nonsavanna vegetation, that cover patches of a few hectares to several square kilometers, are associated with particular sandy or calcareous soils. Soil characteristics and soil available moisture (often related to topography) are the major determinants of the prevailing type of vegetation (see Chap. 2).

The trees of the cerrado have a characteristic contorted appearance due to the loss of the apical bud after the growing season that is later replaced by a lateral bud. The apical buds of many of these tree species grown under irrigation are not lost and the plants grow erect (Laboriau et al. 1964). The bark is usually thick and often gnarled, supposed to provide fire resistance. In some species with outside diameters of up to 20 cm, up to two thirds of it may be bark (Eiten 1994). The leaves of cerrado species, although showing a great deal of interspecific variation, tend to be larger and more coriaceous that those of nearby wet forests.

In the southwest corner of Brazil near the border with Bolivia and Paraguay is a giant flooded savanna known locally as the *pantanal* (Eiten 1975) which, with an extension of about 400 000 km^2 , is the largest wetland area in the world. The vegetation is a mosaic of rainfall elements associated with the areas of permanent water, and grassland and savanna vegetation floristically and physiognomically related to the cerrado, in the areas away from the watercourses.

Another extensive area of wet and periodically flooded savannas in southern South America are the *llanos de Moxos* (Beck 1983) found to the east in Bolivia and extending to the foot of the Andes (Fig. 1.1). This is a giant flood-savanna with a mosaic of vegetation that varies from grassland to evergreen forest according to the position in the gradient of flooding.

In the northeast of Brazil and adjacent to the cerrado (Fig. 1.1), we find an area with somewhat lower average precipitation than the cerrado, but with a well-documented pattern of years of low precipitation followed by years of average and above average rainfall. This area is known as the *caatinga*, and is not considered to be a savanna by South American ecologists (Sarmiento 1984), but is comparable to areas in Africa that are classed as savannas by African ecologists (Cole 1986).

Another region with a mixed woodland, savanna-parkland, and flooded-savanna type vegetation in southern South America is the so-called *Chaco* region of western Paraguay, eastern Bolivia, and northern Argentina (Fig. 1.1) occupying approximately 800 000 km^2, which has all the characteristics of a tropical savanna, but is excluded by most authors (Sarmiento 1984) on account of the occurrence of winter frost in the southern reaches of the area. Frost, however, also occurs occasionally in the southern fringes of the cerrado, and similar formations in southern Africa are considered savannas by African ecologists (Cole 1986).

North of the equator, savanna vegetation dominates some 500 000 km² of the region that surround the northern and western borders of the Apure-Orinoco river system in western Colombia and central Venezuela, known locally as the *llanos del Orinoco*. This is a grassland area with scattered trees belonging mostly to three species — *Curatella americana, Bowdichia virgilioides*, and *Casearia sylvestris*. In the center of the area a large almost treeless flooded savanna is found, the llanos del Apure.

Other areas of savanna north of the equator are the so-called *Gran Sabana* in Venezuela, characterized by a very unique flora (Sarmiento 1983), the coastal savannas of the Guayanas (Van Donselaar 1965), the savannas of the Rio Branco-Rupununi in Brazil (Sarmiento and Monasterio 1975), and a series of small islands of savanna vegetation of edaphic origin dispersed throughout the Amazonian basin known as the Amazonian campos (Sarmiento 1983) (Fig. 1.1). Further north we find small pockets of savanna vegetation in Central America and in Cuba (where the term *savanna* originated).

Savannas are very extensive in Africa, forming a broad semicircle that has its northern end in the vicinity of the West Coast by the Ivory Coast and its southern base in the Coast from southern Angola to northern Namibia and surrounding the wetter rain forest area of central and western Africa (Fig. 1.2). This more or less continuous belt of savanna and woodland-savanna vegetation differs greatly from area to area in physiognomy and floristics due to very different climatic and soil conditions. In the northern belt (Fig. 1.2) two broad bands of savanna can be distinguished, the Guinea-type savannas that form the transition between savannas and the evergreen moist forests to the south, and the Sudan-type savannas that form the transition between the savannas and the Saharan desert vegetation to the north. The Guinea savannas are savanna-woodlands formed by relatively tall and straight trees, while the Sudan savannas are open xerophytic grasslands with scattered deciduous trees. South of the equator the dominant vegetation is a woodland locally known as *miombo* with relatively large trees that towards the north forms a transition with the evergreen moist forest, and towards the south is dominated by more xerophytic elements that form a transition with the South African veldt and other semidesert formations (Fig. 1.2). These two regions are joined in eastern Africa by a series of steppes and grasslands such as the well-known Serenguetti plains in Tanzania and Kenya. These last are found typically in regions with less than 700 mm of rainfall, and are dominated by herbaceous vegetation with some shrubs or scattered trees. In old planation surfaces with plintite-dominated soils, a formation of low trees with contorted stems reminiscent of the South American *cerrado* is encountered.

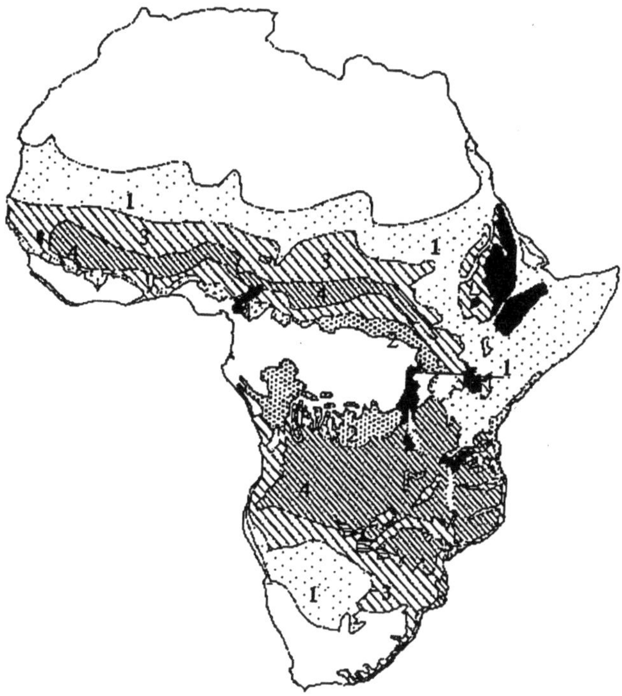

Fig. 1.2. Principal African savanna types. *1* Grassland and shrub savannas; *2* forest-savanna mosaic; *3* tree and shrub savannas; *4* woodland. (After Menaut 1983)

Fig. 1.3. Principal Australian savanna types. *1* Monsoon tallgrass; *2* tropical tallgrass; *3* subtropical tallgrass; *4* midgrass; *5* midgrass on clay soils; *6* tussock grasslands. (After Mott et al. 1985)

Savannas and grasslands are (or were, since much of that area has been transformed by agriculture) fairly extensive in India (Whyte 1968). However, they are thought to be entirely of secondary origin, the result of human activity (Misra 1983). Indian savannas are derived from woodland ecosystems and are found throughout the country. Some secondary savannas and grasslands in the north central region of Rajhastan may have been derived from a primary vegetation that was an open woodland-savanna. In Sri Lanka, savannas and grasslands occupy about a quarter of the country, especially in montane areas, and are most probably also of anthropic origin.

True savannas are rare in the rest of Asia. However, savanna woodlands have been described from Thailand (Blasco 1983) and treeless savannas from Vietnam (Schmid 1974). Secondary savannas have become more extensive as a result of human activity, from Peninsula Malaysia to the Philippines.

Savannas are widespread in Australia in the subhumid regions spreading from south-eastern Queensland across the north of Australia (Fig. 1.3). The Australian savannas can be divided into two more or less distinct types. In the east, temperature is lower, rainfall is on the average higher and the dry season shorter, while in the north, the climate is much warmer and drier and the dry season longer. In both regions there is a gradient of precipitation from the coast towards the interior which is reflected in the vegetation. Within the Australian savannas six major formations have been recognized based on the characteristics of the understory vegetation (Mott et al. 1985): (1) The monsoon tallgrass, tropical tallgrass, and subtropical tallgrass communities that occupy some 500 000 km^2 in a broad belt in the Northern Territory, and northern Queensland, are a low woodland with *Eucalyptus tetradonta, E. dicromophloia*, and *Melaleuca* spp. as dominant species in the woody stratum and a tallgrass layer that may contain *Themeda triandra, Sehima nervosum*, and *Chrysopogon fallax*, in the monsoon tallgrass with tall annual *Sorghum* ssp. on lighter soils; *Heteropogon contortus, Themeda australis*, and *Bothriochloa* spp. in the tropical tallgrass; and *Heteropogon contortus, Schizachyrium fragile*, and *Eriachne* spp. in the subtropical tallgrass accompanied often by *Eucalyptus crebra*, and *E. alba* in the woody stratum; (2) the midgrass savannas and the midgrass savanna on clay soils communities found in three distinct patches in the northern territory and in Queensland that occupy some 100 000 km^2 and are a grassland and open tree formation with *Eucalyptus populnea, E. crebra, E. microneura, E. melanophlioa*, and *Acacia harpophylla* as dominant woody elements according to region, and *Aristida* spp., *Dichantium sericeum, Botriochloa bladhii*, and *Chloris* spp. as the principal grass species; (3) the tussock grasslands. These are extensive interior grasslands of Mitchell grass (*Astrebia* spp.) accompanied by annual species of *Iseilema* spp. and *Dactyloctenium* spp. (Mott et al. 1985).

Physiognomically, savannas vary from open grassland formations to close woodlands dominated by trees 10-20 m tall. The woody stratum may be formed by an almost closed canopy of erect trees such as in the African miombo or some Australian *Eucalyptus tetradonta-E. miniata* communities, or it may be formed by low, contorted trees and shrubs such as in the Brazilian cerrado or African communities dominated by *Uapaca kirkiana.* Almost pure grasslands can be found in areas with high rainfall and a high water table such as in the llanos del Apure, as well as in very dry areas such as in East Africa, or in the tussock grasslands of Australia.

Nevertheless , many ecologists have perceived tropical savannas as a distinct biome (Huntley and Walker 1982; Sarmiento 1984; Frost et al. 1983; Young and Solbrig 1993). Yet it is clear that savannas are very heterogeneous ecosystems that are hard to define. A definition that tries to encompass the diversity of the tropical savanna is as follows (Sarmiento 1984):

> „Savannas are an ecosystem of the warm (lowland) tropics dominated by a herbaceous cover consisting mostly of bunch grasses and sedges that are more than 30 cm in height at the time of maximum activity, and show a clear seasonality in their development, with a period of low activity related to water stress. The savanna may include woody species (shrubs, trees, palm trees), but they never form a continuous cover that parallels the grassy one.“

We will adopt this definition for this book.

1.3 Climate

The principal features of the tropical savanna climate are temperature differences between the warmest and coldest month of the year, overall rainfall, and length of the dry season.

Temperatures depend on the latitudinal location of the savanna and its height above sea level. In savannas located at the margins of the tropics, such as those in southern Africa or in southern South America, as well as some closer to the equator but lying at higher elevations, as is the case in some cerrado localities, extreme low temperatures can approach the freezing mark and differences between mean January and mean July temperatures can be more than 10° C. The latitudinal position of the savanna also determines length of day and distribution of solar radiation (Table 1.1).

Table I.1 Some climatic parameters of savannas

Locality	Country	Lat	Long	Alt	Average Tempe-rature Year	Month with more than 150 mm	Precipi-tation Total	Month with less than 20 mm	Potential evapotrans-priation	No. of Month with ET>P
Brisbane	Australia	-27.28	153.02	41	20	0	1093	0	1056	6
Corrientes	Argentinia	-27.28	-58.50	29	22	0	1268	0	1130	3
Pretoria	South Africa	-25.45	28.14	1460	17	2	746	3	921	9
Bela Vista	Paraguay	-22.06	-56.22	650	22	0	1326	0	1117	6
Campo Grande	Brazil	-20.27	-54.37	566	22	0	1444	0	1085	5
Morondava	Madagascar	-20.17	44.17	10	25	0	777	7	1195	11
Ft. Jameson	Zimbabwe	-19.39	32.41	1256	22	0	1050	6	1071	8
Gwaai	Zimbabwe	-19.17	27.42	1092	21	0	631	6	1081	10
Townsville	Australia	-19.15	146.46	4	24	0	1333	2	1161	10
Vila Pery	Mozambique	-19.08	33.29	732	22	0	1089	5	1065	9
Corumba	Brazil	-19.00	-57.39	127	27	0	1232	1	1239	7
Monte Alegre	Brazil	-18.52	-48.52	756	21	0	1311	3	1047	7
Guaratinga	Brazil	-18.36	-46.38	856	20	0	1448	3	998	6
Catalao	Brazil	-18.10	-47.52	904	22	0	1739	3	1058	6
Burketown	Australia	-17.45	139.33	9	26	0	703	6	1257	10
Paracatu	Brazil	-17.13	-46.52	698	22	0	2851	2	1068	6
Maevatanana	Madagascar	-16.57	46.50	84	27	0	1730	5	1243	8
Goiania	Brazil	-16.41	49.17	729	22	0	1487	3	1064	6
San Ignacio	Bolivia	-16.22	-60.58	335	24	0	1188	1	1151	9
Daly Waters	Australia	-16.16	133.23	211	27	0	667	5	1303	11
Luziania	Brazil	-16.15	-47.46	958	21	0	1475	4	1022	6
Tete	Mozambique	-16.11	33.35	150	26	0	679	7	1248	10
Cacares	Brazil	-16.03	-57.41	117	25	0	1250	2	1181	8
Goias	Brazil	-15.65	-50.08	520	24	0	1786	4	1119	6
Pirenopolis	Brazil	-15.51	-48.58	740	22	0	1695	3	1056	6
Merui	Brazil	-15.43	-51.44	416	23	0	1570	3	1109	7

Locality	Country	Lat	Long	Alt	Average Tempe-rature Year	Month with more than 150 mm	Precipi-tation Total	Month with less than 20 mm	Potential evapotrans-priation	No. of Month with ET>P
Majunja	Madagascar	-15.40	46.21	22	27	0	1587	5	1219	9
Pres. Murtinho	Brazil	-15.38	-53.55	552	22	0	1777	3	1067	6
Zumbo	Mozambique	-15.37	30.27	375	26	0	751	7	1239	10
Cuiaba	Brazil	-15.36	-56.00	172	26	0	1375	3	1187	6
Formosa	Brazil	-15.32	-47.18	912	21	0	1560	4	1037	6
Zomba	Malalawi	-15.23	35.19	957	21	0	1346	5	1051	8
Mongu	Zambia	-15.15	23.10	1053	22	0	969	5	1066	9
Analalava	Madagascar	-14.38	47.46	57	26	0	1912	5	1211	8
Broken Hill	Zambia	-14.24	28.24	1300	21	0	946	5	1028	9
Lilongwe	Malawi	-13.59	33.45	1136	20	2	849	6	992	9
Balovale	Zambia	-13.34	23.06	1091	22	0	975	5	1071	8
Chitembo	Angola	-13.31	16.45	1602	20	0	1135	5	1008	7
Kasempa	Zambia	-13.27	25.30	1483	19	0	1181	5	969	8
Salvador	Brazil	-13.00	-38.30	45	25	0	1866	0	1159	4
Porto Amelia	Mozambique	-12.58	40.30	50	26	0	865	5	1199	9
Nova Lisboa	Angola	-12.48	15.45	1859	19	0	1448	5	962	6
Parana	Brazil	-12.33	-47.47	275	23	0	1338	4	1124	8
Parana	Brazil	-12.33	-47.47	275	23	0	1338	4	1124	8
Lobito	Angola	-12.22	13.12	3	24	0	221	8	1136	12
Luso	Angola	-11.47	19.27	1326	20	0	1193	4	997	6
Ibipetuba	Brazil	-11.01	-44.31	434	24	0	910	5	1134	8
Porto Nacional	Brazil	-10.31	-48.43	237	26	0	1662	3	1182	6
Lindi	Tanzania	-10.00	39.42	41	26	0	897	5	1199	8
Daru	Sierra Leona	-9.04	143.12	8	27	0	2098	0	1222	5
Senn. Madur.	Brazil	-9.04	-68.69	135	25	0	2083	0	1145	4
Sunginge	Angola	-8.46	16.47	689	24	0	1272	4	1145	6
Kilwa	Mozambique	-8.45	39.24	10	27	0	937	4	1216	9
Caungula	Angola	-8.25	18.39	1227	22	0	1722	3	1051	5
Araguaia	Brazil	-8.15	-49.12	1	25	0	1671	3	1172	6

Table 1.1 (cont.)

Locality	Country	Lat	Long	Alt	Average Tempe-rature Year	Month with more than 150 mm	Precipitation Total	Month with less than 20 mm	Potential evapotrans-priation	No. of Month with ET>P
Recife	Brazil	-8.04	-34.53	10	26	0	1757	0	1184	5
Morogoro	Tanzania	-6.57	37.40	633	24	0	859	4	1146	9
Barra Do Cor	Brazil	-5.35	-45.28	82	26	0	1074	4	1193	8
Kigoma	Tanzania	-4.53	29.38	885	24	0	961	4	1118	7
Barumbu	Zaire	1.15	23.29	24	25	0	1798	0	1158	3
Tabou	Ivory Coast	4.25	-7.22	6	26	0	2383	0	1186	4
Santa Elena	Venezuela	4.36	-61.07	998	22	0	1796	0	1034	4
Juba	Sudan	4.52	31.36	457	26	0	982	3	1197	7
San Fernando	Venezuela	5.54	-67.28	73	27	0	1431	3	1230	7
Benin	Benin	6.19	5.37	86	26	0	2076	1	1188	6
Makurdi	Nigeria	7.42	8.35	121	27	0	1405	4	1225	7
Phuket	Thailand	7.58	98.24	3	28	0	2211	0	1244	5
Gambela	Ethiopia	8.15	34.35	1345	27	0	1240	3	1227	7
Raga	Central Africa	8.28	25.41	503	26	0	1146	5	1205	8
Lungi	Sierra Leone	8.37	-13.12	27	26	0	3318	2	1200	6
Barinas	Venezuela	8.38	-70.12	180	26	0	1461	2	1194	6
Villa Bruzual	Venezuela	8.40	-69.18	104	26	0	1493	2	1190	6
Calabozo	Venezuela	8.56	-67.20	119	28	0	1303	4	1235	7
El Baul	Venezuela	8.58	-68.17	102	26	0	1312	3	1I90	8
Yola	Vigeria	9.13	12.29	707	28	0	967	5	1275	8
Malakal	Sudan	9.33	31.39	388	27	0	783	5	1220	9
Minna	Nigeria	9.37	6.32	1059	27	0	1375	5	1242	7
Jos	Nigeria	9.54	8.53	1330	22	0	1404	4	1075	8
Bauchi	Nigeria	10.20	9.50	681	25	0	1097	5	1182	9
Yelwa	Nigeria	10.50	4.45	260	27	0	964	6	1251	9
Kano	Nigeria	12.02	8.32	511	27	0	869	7	1225	10
En Nahud	Sudan	12.42	28.26	564	26	0	423	6	1234	11
Bangalore	India	12.58	77.35	92I	24	0	924	4	1124	9

Locality	Country	Lat	Long	Alt	Average Tempe-rature Year	Month with more than 150 mm	Precipitation Total	Month with less than 20 mm	Potential evapotrans-priation	No. of Month with ET>P
Sokoto	Nigeria	13.01	5.16	1150	28	0	686	7	1315	10
Aranyaprathet	Thailand	13.42	102.35	44	28	0	1524	2	1262	7
Kayes	Mali	14.26	-11.26	47	30	0	826	6	1360	10
Tahoua	Nigeria	14.54	5.15	387	29	0	407	8	1326	11
Menaka	Mali	15.52	2.30	280	30	0	263	8	1404	12
Vishakhapatna	India	17.42	83.18	4	28	0	944	4	I316	9
Poona	India	18.32	73.51	559	25	0	715	5	1210	10
Bombay	India	18.54	72.49	11	27	0	2078	6	1252	9
Akola	India	20.42	77.02	282	27	0	877	6	1352	10
Ahmedabad	India	23.04	72.38	55	27	0	804	8	1362	10

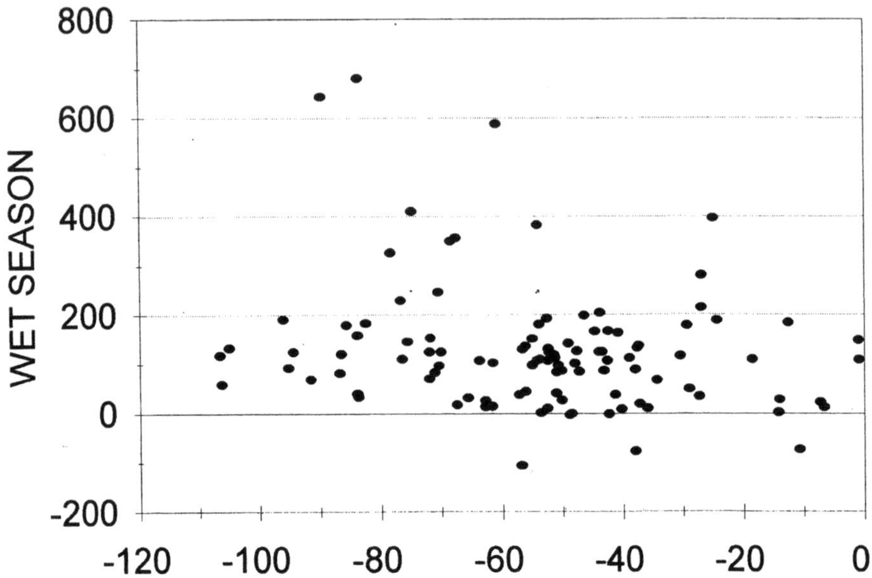

Fig. 1.4. Waterbalance (P-E) in the dry (abcissa) and wet (ordenate) season.

Rainfall varies between 500 mm yr^{-1} and more than 1500 mm yr^{-1} (Table 1.1). Rainfall is a major determinant of the type of vegetation in an area with grasslands prevailing at the low end of the precipitation distribution and woodlands at the high; but rainfall effectiveness is dependent on temperature, and is highly correlated with the length of the dry season. Figure 1.4 shows the relation between precipitation and the length of the dry season for 156 climatic stations in savanna areas of Africa, South America, and Australia. We can see a very close correlation between these two measures, and that savannas span a very wide array of climatic regimes. Figure 1.5 correlates Bailey's index (an index that considers precipitation and temperature) and effective evapotranspiration in the rainy and dry seasons for these same 156 stations, and a few forest and desert localities as well. It can be seen that savanna localities are characterized by a positive water regime (precipitation greater than evaporation) during the rainy season, and a negative one during the dry season. When the water regime becomes positive during the dry season, savannas are replaced by moist forests, and when the water regime tends to be negative throughout the year, xerophytic vegetation prevails.

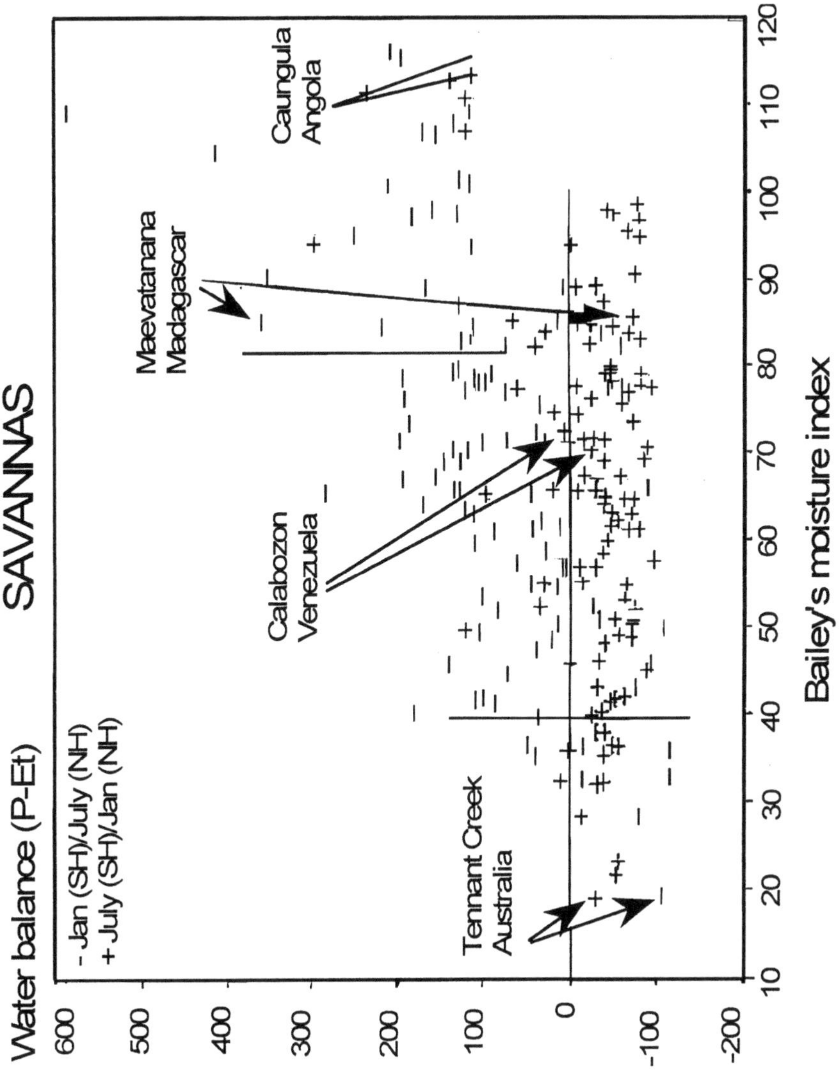

Fig. 1.5. Dry and wet season, precipitation-potential evapotranspiration for 151 savanna and non-savanna localities. Note that with a few exeptions the value is positive during the we5t season, and always negative during the dry season.

In summary, although climate per se does not determine savanna physiognomy, there is an undeniable relation between seasonality and the presence of savannas, and precipitation and woody cover. These relations are not linear, and at both the very wet and very dry end of the spectrum pure or almost pure grasslands may be found.

1.4 Geomorphology and Soils

Africa, South America, Australia, and India were once united, forming the old continent of Gondwana. Gondwanaland begun to break apart in the Jurassic era but the continents did not drift away from each other until the Cretaceous. Many of the geological and soil features of these regions date from the time they formed one continent. The effects of the drifting as well as climatic changes during the Pleistocene created distinct geomorphological features in each continent. Present-day land forms reflect common Gondwanian characteristics as well as the unique individual history of each continent since the Cretaceous.

Tropical savannas occur for the most part on old and weathered surfaces, the result of geomorphological processes over millions of years (Cole 1986). In the regions with high rainfall the landscape is dominated by extensive and very uniform elevated plains surrounded by residual hills. These plains terminate in abrupt escarpments that often lead to lower plains. The area of the escarpments is where the greatest diversity of geomorphological forms is met. The geological history of different continents determines the details of the individual landscapes (Cole 1986). The higher plateau areas are often underlain by lateritic materials that in places have become indurate as a result of exposure. The lower plains often are underlain by calcrete and covered with aeolian deposits. In some regions, such as in the llanos of Venezuela, savannas are found on weathered plains of recent origin (Plio-Pleistocene).

The soils of the major savanna regions reflect their common geological past and their unique recent history. Two major types of soils prevail in savannas: the highly leached, often sandy and lateritic soils known as oxisols or ferralitic soils, and the base-rich montmorillonitic black clay soils known as vertisols. The nutrient status of the soil in tropical savannas is related principally to the age of the sediments. In general savanna soils have low cation exchange capacity, are very low in phosphorus and nitrogen (see Chap. 4), and high in iron and aluminum. The poorest soils (oxisols and ultisols) are those derived from the oldest deposits, since these materials have been subjected to pedogenic processes for prolonged periods of time. Soil nutrient status is a major determinant of vegetation (Chap. 2).

Geology, topography, climate, and drainage affect the physical and chemical characteristics of the soil. Savanna soils vary widely in particle size, structure, profile, and depth, reflecting the interaction of geology, geomorphology, and climate, as well as the influence of topography, the kind of vegetation cover, and animal activity (Young 1976; Montogomery and Askew 1983). Three factors play an important differentiating role in pedogenesis: topography, parent material, and age.

The principal influence that relief has over the ecosystem is on the regulation of drainage, and ultimately over the water balance. In turn, through their action on pedogenesis, the agents that produce the relief indirectly determine the physicochemical characteristics of the soils (Sarmiento 1984) so that relief also translates into the chemical and nutritional characteristics of savanna soils.

Dystrophic savanna soils derived from the weathering of acid crystalline rocks or from ancient sedimentary deposits generally have low reserves of weatherable minerals. The predominance in these soils of 1:1 lattice clays and iron and aluminum oxides results in low effective cation exchange capacity and small amounts of total exchangeable bases, particularly calcium and magnesium (Jones and Wild 1975; Lopes and Cox 1977; Mott et al. 1985). Phosphorus levels are sometimes also very low, and soils rich in sesquioxides have a high capacity for fixing phosphorus. Some highly weatherable soils also have high levels of exchangeable aluminum (Lopes and Cox 1977; Harridasan 1982).

With the exception of extremely acid soils, the amount of organic matter is the main determinant of cation exchange capacity. In wet savannas, high rainfall and an extended wet season favor increased plant production with a consequent input of organic matter into the soil. Because of the almost yearly frequency of fire, the organic matter input is almost exclusively the result of increased below-ground production, since fire effectively mineralizes most of the aerial matter produced (Sanford 1982; Menaut et al. 1985). High temperature and humidity favor microbial activity. However, microbial activity is limited by the low levels of assimilable carbon, high C:N ratios, lignin content, and, in some cases, high amounts of condensed tannins and secondary chemicals.

Microbial activity may be stimulated by root exudates and by water-soluble compounds produced by earthworms (Lavelle et al. 1983; Menaut et al. 1985). According to these authors, these compounds are produced in inverse proportion to their level in the soil, and consequently earthworms and plant roots are supposed to act as regulators of organic matter decomposition in the soil. Because much of the assimilable carbon required by mineralizing bacteria is apparently confined to root exudates and earthworm casts, nutrient release may be highly localized. It has been observed in humid African savannas (Menaut et al. 1985) that absorbing roots often tend to follow earthworm galleries where microbial activity and mineralization are very active. An interesting, but as yet unanswered, question is whether different savanna species differ in the quantity and quality of root exudates they produce, which could have important ecological implications.

The nutrient dynamics of tropical savannas is poorly known. Several authors (Medina 1987, 1993; Sarmiento 1984; Menaut et al. 1985) have summarized existing knowledge on nutrition partitioning between various compartments in the savanna ecosystem and proposed models for the cycling of nitrogen and other elements. The principal conclusion of these studies is that fire represents the principal source of nutrient loss from the system, that internal cycling accounts for the greatest proportion of nutrient fluxes, and that the most important compartment is the organic matter in the soil. The deficit in nitrogen must be covered through rainfall input and free nitrogen fixation.

1.5 Productivity

An accurate appraisal of tropical savanna productivity is essential for understanding the input of organic matter into the ecosystem and the amount and material available for producers and decomposers including those in the soil. Such knowledge is also necessary to understand the potential effect of removal of vegetation as a result of fire, herbivory, and human activity. The new values indicate that the efficiency of light conversion in tropical savanna grasses is higher than previously estimated, which is of economic importance. Finally, an accurate assessment of tropical savanna productivity is indispensable to establish a baseline against which the effects of future changes in global CO_2 levels may be assessed. Savannas differ greatly in productivity, which is not surprising given the great differences in rainfall and soil characteristics.

A number of estimates of the productivity of tropical savanna grasses have been carried out. Most of these studies were made assuming that productivity, the gain of new organic matter by vegetation, approximated the measured increase in above-ground biomass. However, this assumption has proved to be incorrect and has led to underestimation of true productivity by a factor of 2 or 3 (Sarmiento 1984, Long et al. 1989, 1992) due to three reasons: (1) below-ground production can be as high or higher than above-ground biomass; (2) the methods used assumed that death of tissues occurs only after the peak of production has been reached; and (3) the researchers did not consider that different species reach their peak of production at different times. Recent studies (Table 1.2) in tropical grasslands that took these considerations into account have obtained values that are five to ten times higher than previous studies and approximate the figures for tropical forests.

1.6 Floristic Diversity and Affinities

The floristic composition of the different savannas is quite variable. In the herbaceous layer, two families dominate throughout all the savannas: the Gramineae and Cyperaceae. In the woody vegetation there is no such uniformity (Tables 1.3, 1.4, 1.5) and a different mix of species, genera, and families prevails in each continent, and within continents in each region, due to differences in the physical environment.

The flora of the areas that belonged to the old continent of Gondwana is characterized by certain families that are restricted to them. Some of these families such as Proteaceae, Bombacaceae, and Combretaceae are well represented among the woody elements in the savannas of different continents (Tables 1.3-1.5). This may indicate that a proto-savanna vegetation already existed in Gondwanaland. Other woody elements of the savanna belong to widespread families such as Leguminosae and Myrtaceae, while still others, such as the Velloziaceae and Vochyziaceae in South America, are endemic to one continent.

No complete survey of the flora of any of the main savannas exists. There are, however, a number of partial studies (Aubréville 1950; Hall et al. 1970; Eiten 1972). From these studies it is clear that these systems are very rich in their woody flora composition. The richest is undoubtedly the Brazilian cerrado (Eiten 1972), that has an estimated 10 000 species of woody species (Ratter 1986), and the poorest is the Australian savanna. This may be a reflection of the relative species richness in the three continents. While the cerrado has more species than either African or Australian savannas, it is not as rich in species as the Amazonian forest. African savannas have approximately the same number of species as the more restricted humid forests of Africa (Menaut 1983), and Australian savannas, although poorer in species than American or African ones, have more species than the very restricted wet forests of that continent (Lonsdale and Braithwaite 1991). Most woody species have the C_3 photosynthetic pathway.

The Gramineae and Cyperaceae dominate the herbaceous stratum of all savannas. The number of grass species in any savanna number between 30 and 60 (Medina and Huber 1994) with 6-10 dominant species, while the number of sedge species will be between 15 and 30. Herbaceous Leguminosae are the third most important component in number of species, although their abundance is much less than members of the two graminoid families.

Table 1.2. Estimates of productivity of tropical savavanna grasses

Site	Above ground (g)	Below ground (g)	Total (g)	Rainfall (mm)	Source of data
Fete Ole (Senegal)	82	-	-	209	Singh and Joshi (1972)
Pilani (India)	217	61	278	388	Kumar and Joshi (1972)
Welgevonden (S. Africa)	710	-	-	388	Singh and Joshi (1979)
Nairobi National Park (Kenya)	1071			460	Desmukh (1986)
Serengueti (Tanzania)	520	-	-	~700	Bourliere and Hadley (1970)
Jhansi (India)	1014	524	1538	~700	Shankar et al. (1973)
Kurukshetra (India)	2407	1131	3538	790	Singh and Yadava (1974)
Nairobi National Park (Kenya)	805	1075	1880	800	Long et al. (1992)
Nairobi National Park (Kenya)	3228	-	-	850	Cox and Waithaka (1989)
Rwenzori National Park (Uganda)	730	1572	2302	900	Strugnell and Piggott (1978)
Calabozo (Venezuela)	360	-	-	1022	Medina et al. (1977)
Ban Klong Hoi (Thailand)	1568	468	2036	1077	Long et al. (1992)
Barinas (Venezuela)	604	-	-	1093	Sarmiento and Vera (1979)
Mokawa (Nigeria)	614	-	-	1115	Ohiagu and Wood (1979)
Lamto (Ivory Coast)	498	-	-	1158	Singh and Joshi (1979)
Olokemeji (Nigeria)	680	-	-	1168	Hopkins (1968)
Lamto (Ivory Coast)	830	1320	2150	1300	Menaut and Cesar (1979)
Lamto (Ivory Coast)	1540	2040	5380	1300	Menaut and Cesar (1979)

Table 1.3. Principal trees from the American savannas

Family	Species
Annonaceae	*Annona coriacea* *Xylopia aromatica*
Apocynaceae	*Aspidiosperma dasycarpon* *Aspidiosperma tomentosum* *Hancornia speciosa*
Bignoniaceae	*Jacaranda brasiliana* *Tabebuia caraiba* *Tecoma aurea* *Tecoma caraiba* *Zeyheria digitalis*
Bombacaceae	*Bombax gracilipes*
Burseraceae	*Protium ovatum*
Caryocaraceae	*Caryocar brasiliensis*
Cochlospermaceae	*Cochlospermum regium* *Cochlospermum vitifolium*
Combretaceae	*Terminalia argentea*
Compositae	*Piptocarpha rotundifolia* *Vernonia ferruginea*
Connaraceae	*Connarus fulvus* *Connarus suberosus*
Dilleniaceae	*Curatella americana*
Ebenaceae	*Diospyros hispida*
Eruthroxylaceae	*Erythroxylon suberosum*
Flacourtiaceae	*Casearia sylvestris*
Guttiferae	*Kielmeyera coriacea* *Caraipa llanorum*

Table 1.3. (cont.)

Family	Species
Loganiaceae	*Antonia ovata*
Leguminosae	*Hymeneae stigonocarpa* *Machaerium angustifolium* *Piptadenia macrocarpa* *Piptadenia peregrina* *Platypodium elegans* *Pterodon pubescens* *Sclerolobium paniculatum* *Stryphonodendron barbatimao*
Loganiaceae	*Strychnos pseudoquina*
Melastomaceae	*Miconia argentea*
Myrsinaceae	*Rapanea guianensis*
Myrtaceae	*Eugenia dysenterica* *Myrcia tomentosa*
Opiliaceae	*Agonandra brasiliensis*
Palmae	*Acrocomia sclerocarpa* *Attalea exigua* *Astrocarum campestre* *Butia leiospatha* *Copernicia prunifera* *Mauritzia Syagrus campestris* *Syagrus comosus*
Proteaceae	*Roupala acuminata* *Roupala complicata* *Roupala heterophylla*
Rhamnaceae	*Rhamnidium eleiocarpum*
Rosaceae	*Hirtella glandulosa*
Rubiaceae	*Genipa americana* *Platycarpum orinocense*
Sapindaceae	*Magonia pubescens*
Symplocaceae	*Symplocos lanceolata*
Velloziaceae	*Vellozia flavicans*

Table 1.4. Principal trees from the African savannas

Family	Species
Anacardiaceae	*Sclerocarya birrea*
Araliaceae	*Cussonia barteri*
Bignoniaceae	*Daniella oliverii*
Bombacaceae	*Adansonia digitata* *Bombax costatum*
Burseraceae	*Commiphora karibensis* *Commiphora africana*
Combretaceae	*Anogeissus leiocarpus* *Terminalia sericeae* *Terminalia glaucescens* *Combretum apiculatum*
Dipterocarpaceae	*Monotes glaber* *Monotes kerstingii*
Euphorbiaceae	*Hymenocardia acida* *Uapaca kirkiana* *Uapaca nitida* *Uapaca robynsii* *Uapaca togoensis*

Table1.4. (cont.)

Family	Species
Leguminosae	*Acacia seyal* *Acacia radiata* *Acacia nilotica* *Acacia sieberiana* *Baikiaea plurijuga* *Bauhinia* *Brachystegia alienii* *Brachystegia boehmii* *Brachystegia floribunda* *Brachystegia longifolia* *Brachystegia spiciformis* *Brachystegia tamarinoides* *Brachystegia taxifolia* *Burkeae africana* *Colophospermum mopane* *Cryptosepalum pseudotaxis* *Guibourtia conjugata* *Isoberlinia doka* *Isoberlinia globiflora* *Isoberlinia paniculata* *Isoberlinia tomentosa* *Julberlandia globiflora* *Julberlandia paniculata* *Parkia clappertoniana* *Pterocarpus brenanii* *Pterocarpus angolensis*
Melastomaceae	*Afzelia africana*
Ochnaceae	*Lophira lanceolata*
Proteaceae	*Faurea saligna* *Faurea speciosa* *Protea caffra* *Protea elliottii*
Rosaceae	*Parinari mobola* *Parinari curatellifolia*
Sapotaceae	*Butyrospermum paradoxum*
Sterculiaceae	*Sterculia*

Table 1.5. Principal trees from the Australian savannas

Family	Species
Bombacaceae	*Adansonia gregorii*
Cochlospermaceae	*Cochlospermum fraseri*
Combretaceae	*Terminalia canescens*
	Terminalia volucris
Cycadaceae	*Cycas media*
Euphorbiaceae	*Petalostigma quadriloculare*
Lecythidaceae	*Planchonia careya*
Leguminosae	*Acacia aneura*
	Acacia bidwillii
	Acacia difficillis
	Acacia estrophiolata
	Acacia georginae
	Acacia harpophylla
	Acacia pallida
	Acacia shirleyi
	Acacia senegal
	Bauhinia cunninghamii
	Erythrophloeum chlorostachys
	Erythrophloeum intermedia

Table 1.5. (cont.)

Family	Species
Myoporaceae	*Eremophila mitchellii*
Meliaceae	*Owenia vernicosa*
Myrtaceae	*Calythrix archaeta*
	Eugenia bleeseria
	Eucalyptus alba
	Eucalyptus brevifolia
	Eucalyptus brownii
	Eucalyptus clavigera
	Eucalyptus confertiflora
	Eucalyptus crebra
	Eucalyptus cullenii
	Eucalyptus drepanophylla
	Eucalyptus dichromophloia
	Eucalyptus ferruginea
	Eucalyptus foelscheana
	Eucalyptus grandifolia
	Eucalyptus jensenii
	Eucalyptus leptophleba
	Eucalyptus melanophloia
	Eucalyptus microneura
	Eucalyptus microtheca
	Eucalyptus miniata
	Eucalyptus papuana
	Eucalyptus populnea
	Eucalyptus phoenicia
	Eucalyptus pruinosa
	Eucalyptus shrleyi
	Eucalyptus signata
	Eucalyptus tectifica
	Eucalyptus tetrodonta
	Eucalyptus whitei
	Melaleuca quiquenervia
	Melaleuca symphyocarpa
	Melaleuca viridiflora
	Tristania suaveolens
Palmae	*Livistona humilis*
Pandanaceae	*Pandanus spiralis*
Proteaceae	*Grevillea glauca*
	Grevillea heliosperma
	Grevillea ptedirifolia
	Grevillea parallella
	Grevillea striata
Rubiaceae	*Gardenia megasperma*
Sterculiaceae	*Brachychyton rupestris*

Table 1.6. Principal grass species from the American savannas

Andropogon bicornis
Andropogon hirtiflorus
Andropogon leucostachys
Andropogon selloanus
Andropogon semiberbis
Aristida capillacea
Aristida pallens
Aristida tincta
Axonopus aureus
Axonopus canescens
Axonopus capillaris
Axonopus purpusii
Ctenium chapadense
Diectomis fastigiata
Echinolaena inflexa
Eleusine tristachya
Elyonurus adustus
Elyonorus latiflorus
Hymenachne amplexicaule
Leersia hexandra
Leptocoryphium lanatum
Mesosetum loliforme
Panicum cayense
Panicum laxum
Panicum olyroides
Panicum spectabile
Parathenia prostata
Paspalum acuminatus
Paspalum densum
Paspalum carinatum
Paspalum gardnerianum
Paspalum fasciculatum
Paspalum pectinatum
Paspalum plicatulum
Paspalum pulchelum
Paspalum repens
Paspalum virgatum
Setaria geniculata
Setaria gracilis
Sporobulus cubensis
Thrasya paspaloides
Thrasya petrosa
Trachypogon canescens
Trachypogon montufari
Trachypogon plumosus
Trachypogon vestitus
Tristachya chrysothrix
Tristachya leiostachya

Table 1.7. Principal grass species from the African savannas

Aristida spp
Andropogon
Brachyaria obtusiflora
Chloris gayana
Ctenium newtonii
Cymbopogon afronardus
Cymbopogon plurinodis
Cynodon dactylon
Diplachne fusca
Echinochloa pyramidalis
Eragrostis pallens
Eragrostis superba
Heteropogon contortus
Hyparrhenia cymbaria
Hyparrhenia diplandra
Hyparrhenia dissoluta
Hyparrhenia filipendula
Hyparrhenia rufa
Imperata cylindrica
Loudetia demeusei
Loudetia kagerensis
Loudetia simplex
Monocymbium ceresiiforme
Oryza barthii
Pennisetum clandestinum
Pennisetum purpureum
Schizachyrium semiberbe
Schmidta bulbosa
Setaria
Sporobulus robustus
Sporobulus spicatus
Stipagrostis uniplumis
Themeda triandra
Trachypogon capensis
Vossia cuspidata

Table 1.8. Principal grass species from the Australian savannas

Alloteropsis semialata
Aristida contorta
Aristida latifolius
Aristida leptopda
Arundinella nepalensis
Astrebia elymoides
Astrebia lappacea
Astrebia pectinata
Astrebia squarrosa
Bothriochloa intermedia
Chinachne cyathopoda
Chloris
Chrysopogon fallax
Chrysopogon pallidus
Dichanthium fecundum
Dichanthium sericeum
Elyonurus citreus
Eragrostis
Eriachne arenacea
Eriachne obtusa
Eriachne stipacea
Eriachne trisecta
Eulalia fulva
Heteropogon contortus
Heteropogon triticeus
Imperata cylindrica
Iseilema vaginiflorum
Oryza
Panicum mindanenese
Paspalidium
Plectrachne pungens
Pseudopogonatherum irritans
Schizachyrium fragile
Sehima nervosum
Setaria surgens
Sorghum intrans
Sorghum plumosus
Spinifex hirsutus
Sporobulus virginicus
Themeda australis
Themeda aveacea
Triodia irritans
Triodia inutilis

There are many genera of grasses and sedges in common between savannas (Tables 1.6, 1.7, 1.8) although there are no common species, except in those cases of the recent naturalization of African grasses in American savannas (Chap. 6). With very few exceptions, all dominant grasses have the C_4 syndrom in seasonal savannas, and only in very wet environments do C_3 grass species become abundant. The Andropogoneae, Paniceae and Chlorideae are represented by the largest number of species (Johnson and Tothill 1985). Less frequent are the Oryzeae and the Arundinellae. The Andropogoneae are the dominant tribe in the core savanna areas, while the Panicoid dominate in the drier environments, and the Oryzae in the very wet and swamp savannas. The panicoids appear to have the greatest ecological range (Medina and Huber 1994). C_4 grasses can be further divided into two groups: those that use NADP-malic enzyme (malate formers) and those that possess PEP-carboxylase (aspartate formers). It has been shown (Ellis et al. 1980; Medina and Huber 1994) that malate formers dominate in humid savannas and aspartate formers in dry savannas.

1.7 Fauna

Our knowledge of savanna fauna is quite spotty. Only vertebrates are adequately known. The vertebrate fauna of savannas in different continents are totally unrelated, while they show great similarities within continents. African savannas are characterized by the presence of a large number of large herbivores which are absent from American and Australian savannas, although large herbivores might have been present in the American savannas until the end of the Pleistocene. The bird fauna of the wet savannas is extremely rich in all three continents. The faunistic diversity adds one more element to the variety of savanna ecosystems.

1.8 Uses

Savannas are not very productive on account of their poor soils and seasonal climate. Agriculture is practiced in areas with better soils and rainfall over 700 mm. Shifting cultivation has been practiced in Africa for thousands of years. With the increase of population, fallow periods have become shorter, leading often to land degradation (Young and Solbrig 1993). Commercial agriculture directed primarily at the growing of soybeans has become widespread in certain areas of the Brazilian cerrado (Klink et al. 1993). This requires large capital expenditures in land preparation and phosphorus fertilization as well as in yearly nitrogen fertilizations. There is some question whether this type of agriculture is sustainable.

Cattle raising by nomad pastoralists is at least 6000 years old in Africa and India. This system of land use was highly sustainable until disrupted by European colonization. It is still practiced extensively all over Africa, but increased population pressures, conflicts with agriculturalists, and civil wars and international disputes, put in question the future viability of this type of land use.

Extensive commercial ranching is the preferred use of the savanna in Australia and South America. Capital investments and improvements in this type of exploitation are minimal, limited mostly to improving the quality of the herds. The use of fire as a management tool to improve the quality of the grass cover is widespread and can lead to a reduction of the woody cover and to land degradation if the animal load is too high, which is usually the case (Young and Solbrig 1993).

More intensive modern ranching is practiced in areas close to markets and in the better soils. The principal problem in savanna ranching is to provide enough high quality fodder during the dry season. In intensive ranching, natural pastures are replaced partially or totally by planted pastures with a high proportion of cultivated Leguminosae. Such pastures cannot be maintained without fertilization, especially phosphorus fertilization if legumes are part of the pasture, and may also require irrigation during the dry season; but if irrigation is feasible, agriculture is usually a more lucrative alternative.

Savanna regions are for the most part areas of low human density especially in the Americas and in Australia. This is changing somewhat in Brazil with the moving of the capital to the city of Brasilia, situated in the very middle of the cerrado region.

1.9 Summary

In summary, the principal characteristic of tropical savannas is their diversity. Diversity in the climate, in soil types, in topography, in vegetation, flora, fauna, and land use. This diversity makes every savanna type unique and distinct. The role of diversity in the function of this ecosystem will be the subject of the next eleven chapters.

References

Aubréville A (1950) Flore forestière soudano-guinéenne. Soc Ed Geogr Mar Coll, Paris, 523 pp
Beck S (1983) Vegetationsökologische Grundlagen der Viehwirtschaft in den Überschwemmungs-Savannen des Rio Yacuma (Departamento Beni, Bolivien). Diss Bot 80:1-186
Blasco F (1983) The transition from open forest to savanna in continental southeat Asia. In: Bourlière F (ed) Tropical savannas, vol 13 Elsevier, Amsterdam, pp 167-182
Boulière F, Hadley M (1970) The ecology of tropical savannas. Ann Rev Ecol and Syst 1:125-52
Braithwaite RW (1990) Australia's unique biota: implications for ecological processes. J Biogeogr 17:347-354
Bourlière F, Hadley M (1983) Present-day savannas: an overview. In: Bourlière F (ed) Tropical savannas, vol 13. Elsevier, Amsterdam, pp 1-17
Cole MM (1986) The savannas. Biogeography and geobotany. Academic Press, London, 438 pp
Cox GW, Waithaka JM (1989) Estimating above-ground net production and grazing harvest by wildlife on tropical grassland range. Oikos 54:60-6
Deshmukh IK (1986) Primary production of a grassland in Nairobi National Park, Kenya. J App Ecol 23:115-23
Eiten G (1972) The cerrado vegetation of Brazil. Bot Rev 38:201-341
Eiten G (1975) The vegetation of the Serra do Roncador. Biotropica 7:112-135
Eiten G (1994) Vegetacao do cerrado. In: Novaes Pinto M (ed) Cerrado. Caracterizaçåo, ocupaçåo e perspectivas. Editora Univ Brasilia, Brasilia, pp 17-73
Ellis RP, Vogel JC, Fuls A (1980) Photosynthetic patways and the geographical distribution of grasses in South West Africa/Namibia. S Afr J 76:307-314
Frost PGH, Menaut JC, Medina E, Solbrig OT, Swift M, Walker B (1983) Responses of savannas to stress and disturbance. Biol Int Spec Issue 10
Hall N, Johnston RD, Chippendale GM (1970) Forest trees of Australia. Dep Nat Dev, Forestry and Timber Bureau, Canberra, ACT, 334 pp
Harridassan M (1982) Aluminium accumulation by some carrado native species in central Brazil. Plant Soil 65:265-273
Hopkins D (1968) Vegetation of the Olokemeji forest reserve, Nigeria V. The vegetation in the savanna site with special reference to its seasonal changes. J Ecol 56:97-115
Huber O (1979) The ecological and phytogeographical significance of the actual savanna vegetation in the Amazon territory of Venezuela. 5th Int Symp Assoc Trop Biol Macuto
Huntley BJ, Walker BH (1982) Ecology of tropical savannas. Springer, Berlin Heidelberg New York
Johnson RW, Tothill JC (1985) Definition and broad geographical outline of savanna lands. In: Tothill JC, Mott JJ (eds) Ecology and management of the world's savannas. Austr Acad Sci, Canberra, pp 1-13
Jones MS, Wild A (1975) Soils of the West African Savanna. Comm Agric Bur, Tech Comm 55, 246 pp
Klink CA, Moreira AG, Solbrig OT (1993) Ecological impact of agricultural development in the Brazilian cerrados. In: Young MD, Solbrig OT (eds) The world's savannas. Economic driving forces, ecological constraints, and policy options for sustainable, land use. Parthenon, Paris, pp 259-282
Kumar A, Joshi MC (1972) The effects of grazing on the structure and productivity of the vegetation near Pilani, Rajasthan, India. J Ecol 60:665-74
Laboriau LG, Marquez Valio IGF, Heringer EP (1964) Sobre o sistema reproductivo de plantas do cerrado. An Acad Bras Cienc 36:449-464
Lavelle P, Sow B, Schaefer R (1983) The geophagous earthworm, community in the Lamto savanna (Ivory Coast): niche partitioning and utilization of soil nutritive resources. In:

Didal D (ed) Soil biology as related to land use practices. U S Environ Protect Agency, Washington DC, pp 653-672
Long SP, Garcia-Moya E, Imbamba SK, Kamnalrut A, Piedade MTF, Scurlock JMO, Shen YK, Hall DO (1989) Primary productivity of natural grass ecosystems of the tropics: a reappraisal. Plant Soil 115:155-166
Long SP, Jones MB, Roberts MJ (1992) Primary productivity of grass ecosystems of the tropics and subtropics. Chapman and Hall, London
Lonsdale WM, Braithwaite RW (1991) Assessing the effects of fire on vegetation in tropical savannas. Austr J Ecol 16:363-74
Lopes AS, Cox FR (1977) A survey of the fertility status of surface soils under 'cerrado' vegetation in Brazil. J Soil Sci Soc Am 41:742-74
Medina E (1987) Nutrients. Requirements, conservation and cycles of nutrients in the herbaceous layer. In: Walker BH (ed) Determinants of tropical savannas. IUBS, Paris, pp 39-66
Medina E (1993) Mineral nutrition: tropical savannas. Progr Bot 54:237-253
Medina E, Huber O (1994) The role of biodiversity in the functioning of savanna ecosystems. In: Solbrig OT, van Emden HM, van Oordt PGWJ (eds) Biodiversity and global change. CAB Int, Wallingsford, pp 141-160
Medina E, Mendoza A, Montes R (1977). Balance nutricional y producción de materia orgánica en las sabanas de *Trachypogon* de Calabozo, Venezuela. Bol Soc Venez Cienc Nat 134:101-120
Menaut JC (1983) The Vegetation of African Savannas. In: Bourlière F (ed) Tropical savannas, vol 13. Elsevier, Amsterdam, pp 109-150
Menaut JC, Barbault R, Lavelle P, Lepage M (1985) African savannas: biological systems of humification and mineralization. In: Tothil JC, Mott JJ (eds) Ecology and management of the world's savannas. Austr Acad Sci, Canberra, pp 14-33
Menaut JC, Cesor J (1979) Structure and primary productivity of Lamto Savannas, Ivory Coast. Ecol 60:1197-210
Misra R (1983) Indian Savannas. In: Bourlière F (ed) Tropical savannas, vol 13. Elsevier, Amsterdam, pp 151-166
Montgomery RF, Askew GP (1983) Soils of tropical savannas. In: Bourlière F (ed) Tropical savannas. Elsevier, Amsterdam, pp 63-78
Mott JJ, Williams J, Andrew MH, Gillison AN (1985) Australian savanna ecosystems. In: Tothill JC, Mott JJ (eds) Ecology and management of the world's savannas. Austr Acad Sci, Canberra, pp 56-82
Novaes Pinto M (1994) Cerrado. Caracterizaçåo, ocupaçåo e perspectivas. Editora Univ Brasilia, Brasilia
Ohiagu CE, Wood TG (1979) Grass production and decomposition in Southern Guinea Savanna, Nigeria. Oecologia 40:155-65
Ratter JA (1986) Notas sobre a vegetacao de fazenda Agua Limpa (Brasilia, DF) com una chave para os generos lenhosos de dicotyledoneas do Cerrado. Editora Univ Brasilia, Brasilia
Sanford WW (1982) The effect of seasonal burning: a review. In: Sanford WW, Yefusu HM, Ayensu JSO (eds) Nigerian savanna. Kainji Res Inst, New Bussa, Nigeria, pp 160-188
Sarmiento G (1983) The savannas of tropical America. In: Bourlière F (ed) Tropical savannas. Ecosystems of the world, vol 13. Elsevier, Amsterdam, pp 246-288
Sarmiento G (1984) The ecology of neotropical savannas. Harvard Univ Press, Cambridge
Sarmiento G, Monasterio M (1975) A critical consideration of the environmental conditions associated with the occurrence of savanna ecosystems in tropical America. In: Medina E, Golley G (eds) Tropical ecological systems. Springer, Berlin Heidelberg New York pp 223-250
Sarmiento G, Vera M (1979) Composición, estructura, biomasa y producción de diferentes sabanas en los Llanos de Venezuela. Bol Soc Venez Cienc Nat 136:5-41

Schmid M (1974) Végétation du Vietnam. Le massif sub-Annamitique et les regions limitrophes. Memoires OSTROM 74:1-243
Shānkar V, Shankarnavan KA, Rai P (1973) Primary productivity, energetics and nutrient cycling in *Sehima-Heteropogon* grassland. I. Seasonal variations in composition, standing crop and net production. Trop Ecol 14:238-51
Singh JS, Joshi MC (1979) Primary production. In: Coupland RT (ed) Grassland ecosystems of the world. IBP Vol 18, Cambridge Univ Press, Cambridge pp 179-225
Singh JS, Yadava PS (1974) Seasonal variation in composition, plant biomass, and net primary productivity of a tropical grassland at Kurukshetra, India. Ecol Monogr 44:351-76
Solbrig OT (1993) Ecological constraints to savanna land use. In: Young MD, Solbring OT (eds) The world's savannas. Parthenon, Paris, pp 21-48
Strugnell RG, Piggott CD (1978) Biomass, shoot-production and grazing of two grasslands in the Rwenzori National Park, Uganda. J Ecol 66:73-96
Van Donselaar J (1965) An ecological and phytogeographical study of northern Surinam savannas. Wentia 14:1-163
Whyte R O (1968) Grasslands of the monsoon. Faber and Faber, London
Young A (1976) Tropical soils and soil survey. Cambridge Univ Press, Cambridge
Young MD, Solbrig OT (1993) The world's savannas. Economic driving forces, ecological constraints, and policy options for sustainable land use. Parthenon, Paris

2 Determinants of Tropical Savannas

2 Determinants of Tropical Savannas

Otto T. Solbrig, Ernesto Medina, and Juan F. Silva

2.1 Introduction

Tropical savannas, defined as ecosystems formed by a continuous layer of graminoids (grasses and sedges) with a discontinuous layer of trees and/or shrubs, are the most common vegetation type (physiognomy) in the tropics. Tropical savannas are found over a wide range of conditions: rainfall from approximately 200 mm to 1500 mm a year, temperature from subtropical regimes such as the South American Chaco and the South-African savannas with temperature seasonality and cold-month average temperatures below 10 °C, to low-latitude savannas with no temperature seasonality, and soils from volcanic soils such as in parts of the Serengueti plains in Tanzania to dystrophic soils such as in the Brazilian cerrados. The one constant climatic characteristic of tropical savannas is rainfall seasonality. Yet the duration of the dry season can vary from 3 to 9 months, with a mode of 5 to 7 months.

Savannas can be subdivided into a number of savanna types (Table 2.1; Sarmiento 1984) based on rainfall, seasonality characteristics, and density of woody vegetation. These types are not always persistent in time, and natural and anthropogenically induced changes in climate, in nutrients, in fire regime, and in herbivory, can displace the borders of the areas occupied by the various types of savanna vegetation, and the borders with other types of vegetation: humid forests and semideserts. A good example is provided by the border between the Brazilian savanna known as the cerrado and the tropical forest. It is well documented (van der Hammen 1989; Furley et al. 1992) that during the Pleistocene dramatic expansions and shrinkage in the extent of the cerrado took place.

Ecological Studies, Vol. 121
Solbrig, Medina and Silva (eds) Biodiversity and Savanna Ecosystem Processes

Table 2.1. Physiognomic types of savannas (Sarmiento 1984)

1. Savannas without woody species taller than the herbaceous stratum: *grass savannas* or *grasslands.*
2. Savannas with low (less than 8 m) woody species forming a more or less open stratum.
   a) Shrub and or trees isolated in groups; total cover of woody species less than 2%: *tree* and *shrub savanna.*
   b)Total tree/shrub cover between 2 and 15%: *savanna woodland, wooded grassland,* or *bush savanna.*
   c) Tree cover higher than 15%: *woodland.*
3. Savannas with trees over 8 m.
   a) Isolated trees with less than 2% cover: *tall tree savanna*
   b) Tree cover 2 to 15%: *tall savanna woodland.*
   c) Tree cover 15 to 30%: *tall wooded grassland.*
   d) Tree cover above 30%: *tall woodland.*
4. Savannas with tall trees in small groups: *park savanna.*
5. Mosaic of savanna units and forests: *park.*

Savannas from different continents share very few Linnaean species particularly among the woody elements. The invasion of American and Australian savannas by African grasses is a recent phenomenon of human origin. Within an area, however, different savanna types often share common species (Sarmiento 1984; Cole 1986; Medina and Huber 1992). From a floristic point of view, savannas from different continents are quite distinct and show more similarities with other local vegetation types than with savannas in other continents. So, for example, the phylogenetic affinities of the flora of the Brazilian savannas known as *cerrado* are with the Amazonian flora, rather than with the flora of West Africa; however, physiognomically, the cerrado is more similar to the savannas of West Africa than to the Amazonian forest. In turn, African savanna vegetation types are floristically more related among themselves than they are with savanna vegetation in other continents. The conclusion is inevitable: the biota of different savannas are the result of convergent evolution from different floristic and faunistic stocks. As such they provide an interesting puzzle for the evolutionary ecologist: to identify the selective forces that create this unique and widespread tropical physiognomy.

Savanna ecologists have emphasized the similarities rather than the differences in savanna ecosystems. One such approach was the RSSD (Responses of Savannas to Stress and Disturbance) program of the Decade of the Tropics sponsored by IUBS, that developed a set of hypotheses to explain the function of tropical savannas (Frost et al. 1986; Walker 1987; Sarmiento 1990; Werner 1991). They postulated four principal selective forces – which were called determinants – to explain some of the common features and differences in savanna structure and function. These are:

(1) plant-available moisture (PAM), (2) plant-available nutrients (PAN), (3) fire, and (4) herbivory. These determinants are predicted to interact at all ecological scales from landscapes to local patches, but their relative importance differs with scale (Medina and Silva 1990; Solbrig 1991a).

According to the RSSD model, PAM and PAN are the principal determinants of savanna structure at the higher scales. They circumscribe what was called the PAM-AN plane. Where PAM and PAN have high values, woody elements dominate, and as PAM and/or PAN increase eventually the savanna gives way to a moist forest. When PAM and/or PAN have very low values, xerophytic elements prevail, and if the values of the PAM-AN plane get very low, the savanna is replaced by a semidesert. Between these extremes the gamut of savanna types is encountered. To a limited extent PAM and PAN compensate each other: low humidity regimes with relatively high nutrient levels, such as the Serengueti have a savanna and not a semidesert vegetation; likewise, areas with high rainfall but low nutrients, such as the American Llanos del Orinoco and the West African Guinea savannas in the Lamto area, have a savanna rather than a forest vegetation. Within savanna ecosystems, the local effects of the patchy distribution of soil types and topographic features modify PAM and PAN, and together with fire and herbivory, are supposed to determine the density of the tree layer, the productivity of the system, and the rates of nutrient and water flow through the system.

Savannas are very heterogeneous systems in which small forests along streams and moist areas and scattered trees are interspersed in a sea of graminoid-dominated vegetation. So, for example, in the savannas of the Orinoco, the density of the woody vegetation varies with soil depth (PAM) and with the age of the deposits (PAN) (Silva and Sarmiento 1976). Experimental exclusion of fire or herbivores (mostly introduced cattle) produces significant changes in vegetation structure, primarily an increase in the density of woody elements, but also changes in species composition (Braithwaite and Estbergs 1985; Frost and Robertson 1987; Lonsdale and Braithwaite 1991; San Jose and Fariñas 1991; Moreira 1992).

The RSSD program addressed primarily questions regarding the function of savanna ecosystems and ignored somewhat, but not entirely, the behavior of individual species. Yet physical factors of climate and geology – such as rainfall, temperature, soil structure, and soil nutrients – operating on individual organisms, as well as interactions between organisms, constitute the evolutionary forces that shape the characteristics of ecosystems. System properties such as productivity, structure, and resilience are not under direct selection, but are modified as a result of changes in species populations and their properties. All ecosystem properties are the result of a particular mix of species in time and space possessing a given set of characteristics. What is so intriguing is that tropical savannas in different continents, when growing under similar values of PAM and PAN, exhibit very similar ecosystem properties in spite of being composed of different Linnaean species. The question then is not whether ecosystem properties are modified when species change, but how and by how much. The objective of this book is to address explicitly the role of species diversity rather than environmental factors in the function of tropical savannas.

In assessing the role of biodiversity on the function of savanna ecosystems, our first task is to define clearly what we mean by biodiversity in this context. Next, we emphasize the system nature of our approach. Savanna function and structure are two interrelated aspect of the savanna *system*.

2.2 Definition of Biodiversity

"Biodiversity is the property of living systems of being distinct, that is different, unlike. Biological diversity or biodiversity is defined as the property of groups or classes of living entities to be varied. Thus, each class of entity - gene, cell, individual, population and by extension species, communities, and ecosystems - have more than one kind. Diversity is a fundamental property of every living system. Because biological systems are hierarchical, diversity manifests itself at every level of the biological hierarchy, from molecules to ecosystems" (Solbrig 1991b). Here we concern ourselves primarily with species diversity.

The number of different kinds of species living today is enormous, although the exact number is not known with precision. Because taxonomists specializing in different phylogenetic groups have worked in great isolation, not even the exact number of described species is established. They are considered to be approximately 1.6 million. The total number of living species is estimated at between 10 and 30 million, with some computations being as high as 90 million. The exact meaning of all this variation is not known. Two explanations exist, but there is not enough evidence to disprove either. The classical view, held mostly by systematists, physiologists, ecologists, and other organismic biologists, is that processes, such as selection and competition, mold each species to function harmoniously with those surrounding them. In this view, life resembles a machine, a primitive one that does not require very precise parts, but an entity where most parts play designated roles. The other, more recent theory, which is most prevalent among physicists, molecular biologists, geneticists, and mathematicians, sees biodiversity as the manifestation of the dynamics of complex, nonlinear systems, far from equilibrium. For them, life is like a whirlwind in the desert that picks up a lot of dust and materials and arranges it briefly in what for an observer is a defined structure, but where the exact position of each particle is of no significance for the whole.

Neither of these views is entirely correct, nor are they entirely wrong. There exists an enormous body of evidence that indicates that most characteristics of species are adaptive, that is, that they enhance the Darwinian fitness of the individual. Yet there is overwhelming evidence that indicates that not every character is necessarily adaptive, and consequently no species is *optimally* adapted, i.e., possesses the suite of characters that gives it maximal fitness because of the existence of biological processes with a strong random component, such as mutation, and the various environmental factors that produce mortality. Therefore, according to the emphasis put by researchers in their studies, and the scale at which these studies are conducted, the deterministic or the random component of evolution is stressed.

An interesting evolutionary phenomenon is the existence of phylogenetically unrelated biotas with very similar functional properties and a suite of morphologically related species, such as temperate deserts or tropical savannas. The accepted explanation for this situation is that the morphological similarity of the species is the result of convergent evolution under similar selection pressures (Orians and Solbrig 1978). However, since the characteristics of the species in convergent systems are not identical, it is essential that clear predictive hypotheses be formulated indicating in what ways and in which characters convergent species should resemble each other. Otherwise, the fallacy of circularity can be incurred by attributing all similarities to convergent evolution, and all differences to differing phylogenetic histories.

Tropical savannas appear to be convergent in the above sense, even though phylogenetic relations among its species are present. The principal tropical savannas are found in Godwanian continents: South America, Africa, India, and Australia. Consequently, there are some phylogenetic relations especially at the family level and to a lesser extent at the genus level. There are also differences in the degree of phylogenetic similarity between woody and herbaceous elements, the former being less similar in savannas from different continents, the former more. Therefore, when postulating convergent evolution, care must be taken to exclude the effects of common ancestry.

Compared to tropical rain forests, American savannas are poor in species. On the other hand, the African savannas (Menaut 1983) are almost as rich in species as the African rain forests. This may reflect the species poverty of the African rain forest more than the species richness of the African savannas. Australian savannas have more species than adjacent wet forests which are of limited extent in Australia. When compared to temperate grasslands and dry woodlands, tropical savannas are rich in species. Yet there is no complete inventory of the number of species in any tropical savanna. Best known are vascular plants, birds, and mammals; least known are invertebrates, especially non arthropods, fungi, and protists.

The similarity of tropical savannas is predicated primarily on the basis of the vascular vegetation, specifically the presence of a continuous layer of grasses, with or without a discontinuous layer of trees. However, the gramineae are a very stereotyped family of angiosperms, and the characteristics of savanna grass species cannot be necessarily attributed to convergence evolution.

2.3 Savanna Plant Species and Adaptive Strategies

Given that savannas in different continents have very few common species, especially among the woody elements, the similarity in function of different savanna ecosystems implies similarity in function of its different species. Unrelated species with similar functional characteristics are classed as *functional groups*. (Whittaker 1956, 1975; Walter 1973; Grime 1979; Schulze 1982; Körner 1993). Although the concept of functional group can be very useful, it presents serious practical problems. Linnaean species represent groups of similar breeding populations capable of interbreeding, and are consequently nonarbitrary evolutionary units. There is, however, no such nonarbitrary criterion to erect functional groups. The same species may be classed in different functional groups, according to the criterion adopted, since functional groups include all species that share a particular characteristic viewed by the ecologist as important. So, for example, all green plants can be classed as primary producers, all vegetable-consuming organisms as herbivores, and so on. These same species can then be classed as xerophytic or mesophytic according to their water requirements, or they can be classed as drought evaders, or drought endurers, and so on.

A less arbitrary concept to express functional similarity of species is that of *adaptive strategies* (Harper and Ogden 1970; Stearns 1976; Solbrig 1982). The concept of adaptive strategy is based on the notion that species are under constant selective pressure to increase their fitness, and that there are only a limited number of ways in which this can be done on account of developmental constraints (Solbrig 1993). In describing an adaptive strategy, the researcher must postulate the selective forces, as well as the developmental constraints that are responsible for the existence of unrelated species with similar functional characteristics. An adaptive strategy is a hypothesis to explain why a group of species shares certain characteristics. Use of the concept of adaptive strategy forces a greater scientific discipline than arbitrarily classifying species on the basis of shared characteristics, and this explains our preference.

2.4 Species Strategies in Relation to Plant-Available Moisture, and Plant-Available Nutrients

Plants compete for light, water, and nutrients. In savannas, competition is primarily for water and nutrients. There are at least four basic growth forms that have evolved in response to drought: (1) the drought-escaping ephemeral or deciduous perennial; (2) the deep-rooted phreatophyte; (3) the drought-enduring evergreen; and (4) the drought-resistant succulent (Stocker 1968; Barbour 1973; Solbrig 1986; Schulze and Chapin 1987). A drought-escaping plant is one that is active only in the wet season, and survives the period of drought stress as seed or by going dormant; a drought-enduring species is one that continues being physiologically active during the period of drought as long as there is sufficient soil moisture; while a drought-resistant species is a species that has special morphological and physiological adaptations to maintain physiological activity, even under conditions of drought stress. A phreatophyte, on the other hand, escapes the drought by gaining access to the water table. All four strategies are represented in savannas, although the drought-enduring succulent is not an important element. The most important strategies are the deep-rooted phreatophyte represented by the majority of savanna trees and shrubs, the drought-enduring perennial graminoids and the drought--escaping strategy represented primarily by a flora of ephemeral annuals. The distinction between drought-enduring and drought-escaping plants is not as clear in savannas as it is in deserts. Each of these adaptive strategies is characterized by well defined phenological, physiological, and morphological characteristics (Sarmiento et al. 1985; Silva 1987; Medina and Silva 1990; Solbrig et al. 1992). The similarity of the adaptive strategies of savanna plants, particularly among the herbaceous stratum, with those of desert plants, indicates probably that there are no unique savanna-adaptive strategies in relation to drought.

Competition for nutrients is also reflected in the characteristics of savanna species. High transpiration rates by the deep-rooted evergreen woody elements even during the dry season, and high root/shoot ratios have been postulated as primarily mechanisms for gathering nutrients (Sarmiento et al. 1985; Medina 1987). It is possible that the relatively large and very coriaceous leaves of many savanna trees represent a unique characteristic related to their nutrient-gathering strategy. This topic is taken up in detail in Chapter 4.

2.5 Species Diversity, Fire and Herbivory

Plant-available moisture and plant-available nutrients are considered the two principal determinants of savanna function and lead to the concept of the PAM-AN plane, that is theoretically supposed to produce a classification of the world's savannas based upon an ordination of actual sites in relation to these indices. Fire and herbivory were considered to be important in shaping savanna properties at a more local scale (Frost et al. 1986; Medina and Silva 1990; Solbrig 1991a; Teague and Smit 1992).

In relation to both fire and herbivory plant species exhibit two basic adaptive strategies: resistance and escape. Most savanna species are resistant to fire and herbivory and show a series of morphological and phenological traits, such as thick, fire-resistant barks; lignotubers; xerophytic leaves; protected buds; and translocation of nutrients to underground tissues prior to the onset of the dry season when fires occur. These adaptations and the effect of fire on nutrient cycling are well documented (Frost and Robertson 1987). Although fire has probably been present as an important selective factor in tropical savannas for a long time, as attested by the innumerable adaptations of plants to fire, its frequency and intensity has been drastically altered by human presence. Prior to the time when humans started using fire, lightning was the principal source of savanna fires. Some fires are still so started. These fires probably occurred less frequently than fires today, and consequently were of greater intensity. We do not know with precision when humans started using fire as a tool in tropical savannas, but there are remains of hearths as far back as 500 000 years. African savannas have probably been subjected to human-made fires much longer than Australian and American ones, where the human presence is much more recent. No systematic research has been conducted to see whether African species are more resistant to fire than American and Australian ones, but introduced African grasses appear to be more resistant to fire than American species.

There also exists a historical difference between continents in relation to large mammal herbivory. Africa and Asia have had until very recently a widespread and numerous wild fauna of large mammals. Australia never had large herbivorous mammals, and in the Americas the large herbivorous mammals became extinct towards the end of the Pleistocene. It is not known how abundant they were in savanna areas. Cattle was domesticated in the Near East 10 000 years ago and was introduced into African and Asian savannas some 6000 years ago. Its introduction into American and Australian savannas is very recent: in the 16th century in the Americas, only in the 19th in Australia. It is very evident that cattle grazing is a major transforming element in both the American and Australian savannas, and much less so in Africa and Asia. Cattle are also a prime factor in the establishment of introduced African grass species (aside from the fact that their presence is the result of the importation of these species by the cattle industry). The most plausible explanation for the modifying effect of cattle and the greater resistance shown by African species is selection by large herbivores in the Old World savannas. The importance of herbivory is taken up in Chapter 8.

2.6 Conclusions

The basic question to be addressed by the meeting was developed at an initial IUBS/SCOPE/UNESCO meeting held in Petersham, Mass, in 1991. At that meeting the following null hypothesis was developed: "Removal or additions of species (or functional groups, or structural groups, or ecosystem components) that produce changes in spatial configuration of landscape elements will have no significant effect on ecosystem functional properties over a range of time and space scales" (Solbrig 1991b). In the Brasilia workshop, we circumscribed our discussions primarily to species diversity and functional groups, and explored their effect on resistance and resilience to disturbance, water and nutrient flows, fire, and herbivory.

The null hypothesis refers to the addition and subtraction of species. Addition or removal of species from an ecosystem will change the diversity of the system, both its richness and its evenness (Pielou 1975) and by necessity its function. The question is therefore not whether ecosystem properties are modified when species change, but how and by how much. The objective of this book is to address explicitly the role of species diversity rather than environmental factors in the function of tropical savannas.

In the next chapters we present a discussion on the relations between biodiversity and nutrients; biodiversity and water; the impact of alien invasions on biodiversity of the American savannas; the relation between productivity and diversity; the impact of fire and of herbivores; and biodiversity and stability. We end with a presentation of the discussion held in Brasilia (Chap. 10, 11 and 12) and a general summary (Chapter 13).

References

Barbour MG (1973) Desert dogma re-examined: root/shoot, productivity, and plant spacing. Am Midl Nat 89:41-57

Braithwaite RW, Estbergs JA (1985) Fire pattern and woody vegetation trends in the Alligator Rivers region of northern Australia. In: Tothill JC, Mott JJ (eds) Ecology and management of the world's savannas. Austr Acad Sci, Canberra, pp 359-364

Cole MM (1986) The savannas -- biogeography and geobotany. Academic Press, London

Frost PGH, Robertson F (1987) Fire. The ecological effects of fire in savannas. In: Walker BH (ed) Determinants of tropical savannas. IUBS, Paris, pp 93-140

Frost PGH, Medina E, Menaut JC, Solbrig OT, Swift M, Walker BH (1986) Responses of savannas to stress and disturbance. Biol Int Spec Issue 10, IUBS, Paris, pp 1-82

Furley PA, Proctor J, Ratter JA (eds) (1992) Nature and dynamics of forest-savanna boundaries. Chapman and Hall, London

Grime JP (1979) Plant strategies and vegetation processes. Wiley, New York

Harper JL, Ogden J (1970) The reproductive strategies of higher plants. I. The concept of reproductive strategy with special reference to *Senecio vulgaris* L. J Ecol 58:681-698

Körner Ch (1993) Scaling from species to vegetation: the usefulness of functional groups. In: Schulze E-D, Mooney HA (eds) Biodiversity and ecosystem function. Springer, Berlin Heidelberg New York pp 97-116

Lonsdale WM, Braithwaite RW (1991) Assessing the effects of fire on vegetation in tropical savannas. Austr J Ecol 16:363-74

Medina E (1987) Nutrients. Requirements, conservation and cycles of nutrients in the herbaceous layer. In: Walker BH (ed) Determinants of tropical savannas. IUBS, Paris, pp 39-66

Medina E, Huber O (1992) The role of biodiversity in the functioning of savanna ecosystems. In Solbrig OT, van Emden HM, van Oordt PGWJ (eds) Biodiversity and global change. IUBS, Paris, pp 139-158

Medina E, Silva JF (1990) Savannas of northern South America: a steady state regulated by water-fire interactions on a background of low nutrient availability. J Biogeogr 17:403-413

Menaut JC (1983) The vegetation of African savannas. In: Bourlière F (ed) Tropical savannas. Ecosystems of the world 13. Elsevier, Amsterdam, pp 109-150

Moreira A (1992) Fire protection and vegetation dynamics in Brazilian cerrado. PhD Dissertation, Harvard Univ, 201 pp

Orians GH, Solbrig OT (1978) A cost-income model of leaves and roots, with special reference to arid and semi-arid areas. Am Nat 111:677-690

Pielou EC (1975) Ecological diversity. Wiley, New York

San Jose JJ, Fariñas M (1991) Temporal changes in the structure of a *Trachypogon* savanna protected for 25 years. Acta Oecol 12:237-247

Sarmiento G (1984) The ecology of neotropical savannas. Harvard Univ Press, Cambridge

Sarmiento G (ed) (1990) Las sabanas americanas. Aspecto de su biogeografia, ecologia y utilizacion. CIELAT, Mérida, Venezuela

Sarmiento G, Goldstein G, Meinzer F (1985) Adaptive strategies of woody species in tropical savannas. Biol Rev 60:315-355

Schulze E-D (1982) Plant life forms and their carbon, water, and nutrient relations. In: Lange OL, Noble PS, Osmond CO, Ziegler H (eds) Encyclopedia of plant physiology 12B. Springer, Berlin Heidelberg New York, pp 120-148

Schulze E-D, Chapin S III (1987) Plant specialization to environments of different resource availability. In: Schulze E-D, Swolfer H (eds) Potentials and limitations of ecosystem analysis. Springer, Berlin, Heidelberg, New York, pp 120-148

Silva JF (1987) Responses of savannas to stress and disturbance: species dynamics. In: Walker BH (ed) Determinants of tropical savannas. IUBS, Paris, pp 141-156

Silva JF, Sarmiento G (1976) Influencia de factores edaficos en la diferenciacion de las sabanas. Analisis de componentes principales y su interpretacion ecologica. Acta Cient Venez 27:141-147
Solbrig OT (1982) Plant adaptations. In: Bender GL (ed) Reference handbook on the deserts of North America. Greenwood Press, Westwood, pp 419-432
Solbrig OT (1986) Evolution of life-forms in desert plants. In: Polunin N (ed) Ecosystem theory and application. Wiley, Chichester , pp 89-105
Solbrig OT (ed) (1991a) Savanna modeling for global change. Biol Int Spec Issue 24:1-45, IUBS, Paris
Solbrig OT (1991b) Biodiversity. Scientific issues and collaborative research proposals. Mab Digest 9:77, UNESCO, Paris
Solbrig OT (1993) Plant traits and adaptive strategies: their role in ecosystem function. In: Schulze E-D, Mooney HA (eds) Biodiversity and ecosystem function. Springer, Berlin Heidelberg New York pp 97-116
Solbrig OT, Goldstein G, Medina E, Sarmiento G, Silva J (1992) Responses of tropical savannas to stress and disturbance: a research approach. In: Wali MK (ed) Ecosystem rehabilitation. 2. Ecosystem analysis and synthesis. SPB, The Hague, pp 63-73
Stearns SC (1976) Life history tactics: a review of the ideas. Q Rev Biol 51:3-47
Stocker O (1968) Physiological and morphological changes due to water deficiency. Arid Zone Res 15: 63-104
Teague W R, Smit GN (1992) Relations between woody and herbaceous components and the effects of bush-clearing in southern African savannas. Tydskrif Weidingsveren. S Afr 9:60-71
Van der Hammen T (1989) History of the montane forests of the northern Andes. Plant Syst Evol 162:109-114
Walker BH (1987) Determinants of Savannas. IRL, Oxford
Walter H (1973) Die Vegetation der Erde in ökaphysiologischer Betrachtung. Band 1. Die tropischen und subtropischen Zonen. Fischer, Jena
Werner P (Ed) (1991) Savanna ecology and management. Australian perspectives and intercontinental comparisons. Blackwell, Oxford
Whittaker RH (1956) Vegetation of the great smoky mountains. Ecol Monogr 26:1-80
Whittaker RH (1975) Communities and ecosystems. McMillan, New York

3 Biodiversity and Nutrient Relations in Savanna Ecosystems: Interactions Between Primary Producers, Soil Microorganisms, and Soils

3 Biodiversity and Nutrient Relations in Savanna Ecosystems: Interactions Between Primary Producers, Soil Microorganisms, and Soils

Ernesto Medina

3.1 Introduction

Availability of resources (light, water, nutrients) determines the amount of biomass that may be produced in a given environment (Chapin 1980; Tilman 1988; McNaughton 1990). This production of biomass can be brought about by assemblages of primary producers composed of widely different species. Similar environments on different continents attain the same level of organic matter production, and are occupied by more or less equivalent ecosystems in regard to structure and function (Walter 1973). Higher production is the result of efficient trapping of available resources. The capacity for trapping available resources would depend on the ability of the species assemblage to occupy the space: intercept incident light and take up soluble nutrients and water. This ability is regulated by intrinsic factors characteristic of each species such as plant habit, size, specific growth rate, phenology, and physiological requirements. Biological interactions extrinsic to the primary producers are also important and include interspecific competition, organic matter decomposition, and the presence of symbiotic and mutualistic microorganisms.

Higher availability of plant nutrients in a certain environment can result from higher nutrient concentration (i.e., amount of nutrient/unit soil volume), larger soil volume available to plant roots, or both. More nutrients available per unit area of soil allow a larger number of individuals of the same or of different species to be packed in the same available space. Higher density of individuals leads to a more effective light interception and, in

Ecological Studies, Vol. 121
Solbrig, Medina and Silva (eds) Biodiversity and Savanna Ecosystem Processes

absence of water shortage, higher primary productivity. Differences in morphological and/or physiological traits will determine which species can take advantage of this large supply of nutrients.

Savannas are characterized by the coexistence of primary producers of contrasting growth habit: grasses, sedges, shrubs, and trees. These contrasts lead to a stratified utilization of the space available above- and below-ground. Trees are better competitors for light, while grasses are better competitors for nutrients and water (Walter 1973; Walker and Noy-Meier 1982). The competition relationships in any savanna, however, will be determined not only by the ability of primary producers to trap light and nutrients, but also by their tolerance to the impact of fire and herbivory (Medina and Silva 1990; McNaughton 1991).

Differences in size can explain in part the coexistence of different species of primary producers in the same savanna. Smaller species may be more efficient competitors in the exploration of the available soil volume, but are less efficient competitors for light.

This chapter discusses some examples relating biodiversity and nutrient relations of primary producers in savanna ecosystems, focusing on a few life-forms (trees, shrubs, and grasses) for which significant information exists. The large diversity of life-forms and functional groups in savannas (Eiten 1972; Medina and Huber 1992) suggest much more complex relationships deserving special attention in future research.

The chapter is based on two assumptions related to the role of biodiversity in regulating nutrient relations at the ecosystem level in savannas:

1. The degree of resource capture in a given savanna increases with the number of species per unit area. The occurrence of a set of reproducing species in a given environment indicates that resources are available for the cooccurring species at a given time. The cooccurrence does not mean that the composition of the system is stable or at equilibrium. Species displacement can occur due to direct or indirect biotic interactions. Disturbance can lead to a reduction or an increase in the number of species in a given environment depending on the direction that nutrient availability is affected. Environments with too low or too high nutrient availability are frequently associated with low species diversity (Lugo 1988; Tilman 1988).

2. Species composition and relative abundances in a given savanna affect nutrient availability over space and time because of accumulation of certain nutrients resulting in spatial patchiness, and biological properties determined by symbiosis and mutualisms.

3.2 Dominance Spectra of Tropical Savannas: the Significance of Rare Species

Dominance spectra in the grassland layer of tropical savannas show frequently a high degree of dominance by a few grass species (Sarmiento 1983). Nevertheless, detailed inventories of all grassland species show the coexistence of a considerable number of species of primary producers within a given site. Bulla et al. (1984) for example, recorded during the dry season in a series of relatively species-poor savannas in eastern Venezuela a total of 13 species of grasses, 10 species of sedges, 10 species of legumes, and 10 species distributed among other families. Within each of these groups a pattern of strong dominance by a few species was observed (Medina and Huber 1992). From the 43 species recorded, only 2 had frequencies higher than 10%, and 9 had frequencies between 1 and 10%. That is, those species represented with frequencies below 1% were much more numerous than the common species. Similar patterns of dominance have been recorded for the tree and shrub component of savannas. Felfili and da Silva (1993) measured the number of cerrado *sensu stricto* sites in central Brazil and found a number of tree species ranging from 51 to 63. The number of species with Indices of Importance Value (IVI) (sum of relative density, dominance, and frequency) above 10 averaged only 10%, whereas those species with IVIs below 1 averaged 17%.

The question arises immediately, why are there so many „rare" species in these savannas? Our hypothesis is that those species found empty spaces in the community that can be occupied on a more or less permanent basis, making these rare species also characteristic of the primary-producer functional group in these savannas. Competition for nutrients and light is essentially a competition for the occupation of the space available. If a number of individuals of a certain species can become established and reproduce successfully in a community, it means that there is space available, or in other words, that there are resources which are not being utilized by the dominant species in the community. If that is the case, the relationship between diversity and resource capture is straightforward: in a stable community of primary producers, reduction of diversity, represented by complete elimination of one or several species, will necessarily result in resource waste.

3.3 Availability of Nitrogen and Phosphorus and the Role of Symbiotic Microorganisms

Nitrogen availability appears to regulate the productivity of grasslands in many savanna areas (Medina 1987). It has been suggested also that tropical savannas on dystrophic soils are dominated by C_4 species belonging to the malate-forming type, while semiarid grasslands with higher nutrient availability are dominated by C_4 grasses belonging to the aspartate-forming type (Huntley 1982). Long-term fertilization experiments which increased nitrogen availability in native grasslands in South Africa resulted in higher frequencies of aspartate-formers at the expense of malate-formers (O'Connor 1985, Medina and Huber 1992). Even in experiments of shorter duration, with successional dominants in temperate grasslands, it has been shown that the proportion of *Schizachyrium scoparium*, a C_4 species highly efficient in the capture of soil nitrogen, is displaced by *Agropyron repens*, a C_3 grass with high nitrogen requirements, when nitrogen availability is experimentally increased in nitrogen-poor sandy soils (Tilman 1990).

It is also known that different grass species can affect the cycling of nitrogen in cultivated grasslands, because each species has characteristic nitrogen and lignin concentrations in above- and below-ground biomass (Wedin and Tilman 1990). Pure cultures of C_3 and C_4 grasses grown on identical soils depart significantly in their nitrogen mineralization rates after 3 years of cultivation. The C_3 species *Agrostis scabra*, *Agropyron repens* and *Poa pratensis* have significantly higher nitrogen and lower lignin concentrations in the below-ground biomass compared to the C_4 species *Schizachyrium scoparium* and *Andropogon gerardii* (Wedin and Tilman 1990). There are also important differences in concentrations among the species belonging to each group. This is a clear indication that there may be species-specific effects on the cycling of nutrients in a given savanna.

Patches occupied by annual or shrubby legumes in tropical savannas can constitute areas of high nitrogen availability because of their active N_2 fixation. These patches possibly allow the invasion of species with higher nitrogen demand (Izaguirre-Mayoral et al. 1992). That this occurs under natural settings has been shown clearly by the invasion of *Myrica faya*, a shrubby species forming actynomycetal symbiosis, which entered Hawaii at the end of the 19th century, and rapidly invaded pastures and native forests thinned by volcanic eruptions (Vitousek et al. 1987). This species provided a considerable input of biologically fixed nitrogen leading to modification of the patterns of secondary succession in many disturbed habitats in Hawaii. Although processes of secondary succession have not been thoroughly documented in tropical savannas, there are strong indications that trees in general, and legumes in particular, increase soil fertility compared to the soil surrounding grass species (see below).

Phosphorus availability seems to regulate the frequency and productivity of legumes in tropical savannas. Leguminous species, both annual and perennials, in South American savannas appear to be nitrogen-sufficient but are phosphorus-limited, while the dominant grass species are limited by nitrogen, but not by phosphorus (Medina and Bilbao 1991). Under natural conditions, legumes are frequently infected by mycorrhizal fungi, but it is not known how efficient these symbiosis are in providing phosphorus for the host legume (Saif 1986; Cuenca and Lovera 1992). The effect of mycorrhizal fungi on the acquisition of soil phosphorus and the regulation of nodulation in tropical legumes has been shown quite clearly in phosphorus-deficient soils, under both laboratory and field conditions (Crush 1974; Arias et al. 1991). The beneficial effects of mycorrhizal infection for the growth of legumes under field conditions may be therefore very important.

Soil conditions and species composition of the primary producers can also regulate the diversity of mycorrhizal fungi in the soil. In old field succession in temperate grasslands, VAM-fungal biomass density is positively related to higher pH, water-soluble soil carbon, and root biomass, and is inversely related to soil-extractable phosphorus and the frequency of non-host plant species (Johnson et al. 1991). Different grass species growing on soils of contrasting texture also significantly modify the composition of the mycorrhizal fungi community in the soil (Johnson et al. 1992). These facts indicate that biotic interactions are quite important in the regulation of biodiversity, and that changes in species composition of the primary producer's functional group or the species assemblage of mycorrhizal fungi in the soil might have profound influences in the response of the system to stress and disturbance.

3.4 The Effect of Trees on the Physicochemical Characteristics of Soils Underneath

Savanna ecosystems are characterized by a continuous layer of xeromorphic C_4 grasses, interspersed with shrubs and trees which are predominantly deciduous in Africa and evergreen in South America and Australia (Cole 1986). Nutritional requirements of trees and grasses differ substantially, particularly those for potassium and calcium (Medina et al. 1982; Medina 1987). The discrete distribution of trees and grasses, therefore, has to be reflected in the patchiness of soil physicochemical properties.

Kellman (1979) documented higher levels of soil-fertility indicators beneath tree crowns in savannas of Belize. There was an increase in nutrient availability below *Byrsonima crassifolia* but not so under *Pinus caribaea*. Susach (1984) showed a clear enrichment of the soil under the crown of *Curatella americana*, another common evergreen tree in northern

Table 3.1. Changes in the content ot cations (mg/kg air dry soil) in areas under tree canopies and under grasses in Belize and Venezuela. (After Kellman 1979 and Susach 1984)

	Venezuela		Belize		
Cation	*Grassland*	*Curatella*	*Grassland*	*Byrsonima*	*Pinus caribaea*
Ca	93	191	42	148	38
Mg	51	99	26	46	26
K	36	78	25	31	25

South American savannas, compared with the soil under grassland (Table 3.1). In Nigerian savannas (Sudanese woodlands with abundant trees of the genus *Isoberlinia*), Isichei and Muoghalu (1992) found that soil under crowns of large trees was enriched in organic matter, calcium, and potassium, and had a higher cation exchange capacity and pH compared to the soils of the surrounding grasslands. Similar results have been reported by Georgiadis (1989) and Mordelet et al. (1993).

Belsky et al. (1989) showed that organic matter and nitrogen content, phosphorus availability, and exchangeable potassium and calcium were higher under the crowns of the deciduous trees *Acacia tortilis* and *Adansonia digitata* compared to open grasslands in Tsavo National Park in Kenya. The soil differentiation occurs both in high- and low-rainfall areas (Belsky et al. 1993) and influences the nutritional values and the productivity but not the diversity of the grasses growing underneath the crowns (Belsky 1992).

It is clear that in savannas tree density increases available nutrients, water infiltration rate, and soil organic matter compared to open grasslands (Belsky and Amundson 1992). The reasons for this increase in soil fertility under trees are varied (Kellman 1979; Belsky and Amundson 1992): (1) the extensive root system of trees explores large soil volumes and extracts nutrients that are concentrated under the crown through litter fall, (2) trees are used as perches by birds, therefore the area under the crown might be enriched by bird droppings, (3) large mammals that graze in the grassland but rest under the tree shade deposit nutrients in their dung, and (4) trees trap precipitation inputs more efficiently.

It has to be expected that the quality of foliage and litter produced by the trees also influences the process of soil enrichment. Although Belsky et al. (1993) did not observe clear differences in the amount of cations between the soils under *Adansonia digitata* and *Acacia tortilis* trees in Tsavo National Park, the differences between *Byrsonima* and pine in Belize are striking (Table 3.1; Kellman 1979). In addition, the amount of nitrogen accumulated in the herbaceous biomass is much larger under the African leguminous species *Acacia tortilis* than under *Adansonia* and in nearby open grasslands (Table 3.2). Microbial biomass was higher under the tree crowns, but no differences were observed between the species.

Table 3.2. Influence of the canopy tree species *(Acacia tortilis* and *Adansonia digitata)* on the nitrogen content of the biomass of the grass vegetation growing underneath and the soil nitrogen content and microbial biomass in low- and high-rainfall areas of the Tsavo National Park in Kenya. (After Belsky et al. 1993)

	Low-rainfall site			High-rainfall site		
	Acacia	*Adansonia*	Grassland	*Acacia*	*Adansonia*	Grassland
Nitrogen content of biomass (g/m^2)	[illegible].55	4.03	1.16	6.22	2.51	1.71
Nitrogen content in soil (kg/m^3)	1.79	1.70	0.85	1.64	1.62	1.01
Microbial biomass (kg/m^3)	0.34	0.35	0.25	0.93	0.70	0.50

Similar differences may be observed under the crowns of evergreen and deciduous tree species, because nitrogen and phosphorus content per unit leaf weight is higher in deciduous than in evergreen trees in northern South America (Medina 1984). It is also known that trees and grasses differ markedly in their nutritional requirements. For instance, with appropriate soil calcium availability, broad-leaved trees take up significantly more calcium than grasses (Medina 1987). As mentioned previously, perennial legumes have phosphorus requirements well above those of C4 grasses in South American savannas (Medina and Bilbao 1991).

3.5 Differences in Soil Nutrient Availability and Vegetation Gradients in Savannas

The Brazilian Cerrados occupy a large area characterized by the occurrence of open grasslands, shrub- and tree-savannas, and sclerophyllous forests. They constitute a complex pattern of vegetation varying along geomorphological, pedological, and nutritional gradients (Eiten 1972).

Lopes and Cox (1977) surveyed soils under many different vegetation types within the Cerrado area of Brazil, including *campo limpo* (open grassland), *campo sujo* (shrubby savanna), *campo cerrado* (tree-savanna), *cerradão* (sclerophyllous forest), and semideciduous forest. Samples were taken from the upper 30 cm and pooled together according to vegetation types. The data indicated little variation in soil organic matter or exchangeable potassium and aluminum along the gradient. However, the percent of aluminum saturation decreased from the open grassland to the semideciduous forest, while the exchangeable calcium and magnesium increased in the same direction (Fig. 3.1). Extractable phosphorus followed

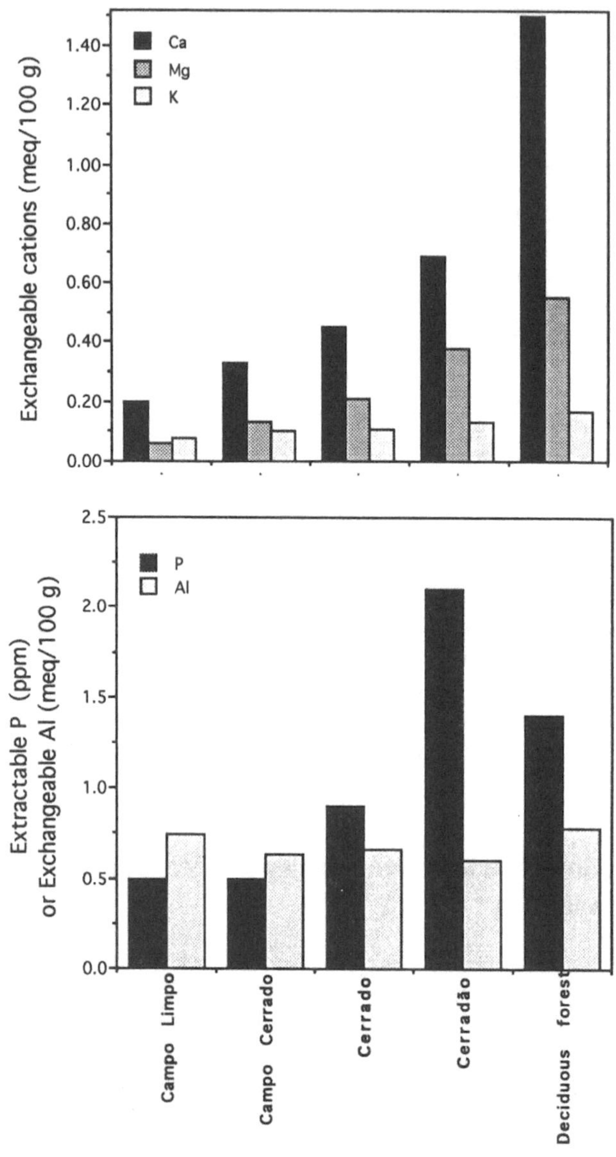

Fig 3.1. Variation in the concentration of cations, Al and extractable phosphorus in superficial soil samples taken under different vegetation types in Brazil. (Data from Lopes and Cox 1977).

the same pattern reported by Belsky and Amundson (1992), because it remained low in open and shrubby grasslands, increased significantly in tree-savanna and was highest in the cerradão and the semideciduous forest (Fig. 3.1). Along these gradients, the number of tree species increased from approximately 30 to 55, and the number of grass species declined from 60 to less than 40 (Goodland and Ferri 1979). The change in both number of tree individuals and tree basal area is quite pronounced, ranging from about less than 1000 individuals and 3 m^2 basal area per ha in the campo sujo plots to more than 3000 individuals with a basal area of more than 30 m^2 per ha in the cerradão plots (Goodland and Pollard 1972). Superimposed on this complex are the ever-increasing human influences determining frequency of fire and deforestation.

The variation in density of woody components has been interpreted as a fertility gradient (Goodland and Pollard 1972), because it appears that availability of nutrients decreases from the open grasslands dominated by relatively low nutrient-demanding grasses to the forest dominated by trees of higher nutrient requirements. Although there is a clear correlation between number of tree species and soil nutritional status (Ratter and Dargie 1992) the pattern is more complex (Furley and Ratter 1988; Oliveira-Filho et al. 1989). In addition, the gradient may also be interpreted as the result of the improvement of soil fertility brought about by the presence of trees.

To decide which of the two explanations is correct is not easy, because in the cerrados, as in the rest of the savanna ecosystems, fire frequency plays a central role, affecting density of woody vegetation. It seems clear that vegetation of seasonal savannas is strictly associated with dystrophic, nutrient-poor soils (Sarmiento 1992). However, there are reports of cerrado vegetation growing on mesotrophic (calcium-rich soils) soils. Haridasan (1992) measured nutrient concentrations in soils and leaves of tree species growing in cerrado *sensu stricto* and cerradão on dystrophic and mesotrophic soils in central Brazil. In each soil group, no differences were detected in the concentration of cations in the soil under, or in the leaves of both vegetation types. The precise evaluation of these differences will, however, require information of the soil bulk density, possibly lower under cerradão than under cerrado, and also on the successional status of the cerrado sites. Striking differences were observed in the concentrations of calcium and magnesium between dystrophic and mesotrophic soils (Table 3.3). Weighted concentrations of calcium, magnesium, and potassium in mesotrophic soils were 200, 20, and 2 times, respectively, higher than in the dystrophic soils. These changes were reflected in the foliage of the trees but in a much lower proportion. Calcium increased 3 times, while magnesium and potassium increased only 1.4 times in mesotrophic, compared to dystrophic soils. These results show that the chemical differences associated with the pedogenetic origin of these two soil types override the differences which might have been produced by the enrichment effect of the trees observed in the majority of savannas.

Table 3.3. Contrasts in cation concentration in soils and tree eaves in woody vegetation in cerrado savannas of central Brazil. (Data from Haridasan 1992)

	On dystrophic soils		On mesotrophic soils		
	Cerrado s.s.	Cerradao	Cerrado s.s.	Cerradao	Rel. difference meso-/dystro
Soil					
Exchangeable cations (weighted average in mg/kg)					
depth (cm)	56	60	45	56	
Ca	9	4	1293	1340	203
Mg	5	3	110	64	22
K	28	20	65	51	2
Plants (mg/g dry matter)					
Ca	3.5	2.3	7.2	10.7	3.1
Mg	2.0	1.8	2.8	2.7	1.4
K	5.0	4.8	6.7	7.5	1.4

3.6 Conclusions

The limited literature reviewed in the present chapter warrant several conclusions of importance to evaluate the role of biodiversity in the nutrient relationships of savanna ecosystems:

1. It seems that structural complexity and diversity of life forms in a savanna ecosystems are associated with the stratified exploitation of resources both above- (light) and below-ground (nutrients and water).
2. Cooccurrence of reproducing species in a given environment reveals the availability of resources being intercepted by those species. In particular, the presence of „rare“, low-frequency species, seems to indicate the inability of the dominant species to capture all the resources available in a given environment.
3. Biotic interactions markedly affect the nutrient availability in a given environment. Rhizobial or actynomycetal symbiosis may significantly modify the availability of nitrogen, while the occurrence of mycorrhizal symbiosis provides a higher phosphorus supply to the host plant.
4. Primary producers and soil microorganisms have strong bi-directional influences. Changes in the composition of one community may result in significant changes in composition of the other, due to variations in the nutritional quality of the substrate, and the availability of certain nutrients.
5. Trees and grasses modify their nutritional environment because of the nutritional quality of organic matter that they produce. These effects are not identical even for species similar to habit and phenology.

Acknowledgments.
To Drs. Joy Belsky (Oregon Natural Resource Council, Portland USA), Augusto Franco (Universidade de Brasilia, Brasil), and Juan Silva (Universidad de los Andes, Mérida, Venezuela) for their valuable comments and criticisms.

References

Arias I, Koomen I, Dodd JC, White RP, Hayman DS (1991) Growth responses of mycorrhizal and non-mycorrhizal tropical forage species to different levels of soil phosphate. Plant Soil 131:253-260

Belsky AJ (1992) Effects of trees on nutritional quality of understorey gramineous forage in tropical savannas. Trop Grassl 26:12-20

Belsky AJ, Amundson RG (1992) Effects of trees on understory vegetation and soils at forest-savanna boundaries. In: Furley PA, Proctor J, Ratter JA (eds) Nature and dynamics of forest-savanna boundaries, chap 17. Chapman & Hall, London, pp 353-366

Belsky AJ, Amundson RG, Duxbury JM, Riha SJ, Ali AR, Mwonga SM (1989) The effects of trees on their physical, chemical, and biological environments in a semi-arid savanna in Kenya. J Appl Ecol 26:1005-1024

Belsky AJ, Mwonga SM, Amundson RG, Duxbury JM, Ali AR (1993) Comparative effects of isolated trees on their under canopy environments in high- and low-rainfall savannas. J Appl Ecol 30:143-155

Bulla L, Sánchez P, Silvio C, Maldonado A, De Sola R, Lira A (1984) Ecosistema sabana. Bases para el diseño de medidas de mitigación y control de las cuencas hidrográficas de los ríos Caris y Pao (Edo. Anzoátegui), vol 1. Inst Zool Trop Fac Cienc Univ Cent Venezuela, Caracas, pp 36-125

Chapin FS III (1980) The mineral nutrition of wild plants. Annu Rev Ecol Syst 11:233-260

Cole MM (1986) The savannas: biogeography and geobotany. Academic Press, London

Crush JR (1974) Plant growth responses to vesicular-arbuscular mycorrhiza. VII. Growth and nodulation of some herbage legumes. New Phytol 73:743-752

Cuenca G, Lovera M (1992) Vesicular-arbuscular mycorrhizae in disturbed and revegetated sites from La Gran Sabana, Venezuela. Can J Bot 70:73-79

Eiten G (1972) The Cerrado vegetation of Brazil. Bot Rev 38:201-341

Felfili JM, da Silva jr MC (1993) A comparative study of cerrado (sensu stricto) vegetation in Central Brazil. J Trop Ecol 9:277-289

Furley PA, Ratter JA (1988) Soil resources and plant communities of the central Brazil cerrado and their development. J Biogeogr 15:97-108

Georgiadis NJ (1989) Microhabitat variation in an African savanna: effects of woody cover and herbivores in Kenya. J Trop Ecol 5:93-108

Goodland R, Ferri MG (1979) Ecologia do Cerrado. Sao Paulo. Univ Sao Paulo, Brazil

Goodland R, Pollard R (1972) The Brazilian Cerrado vegetation: a fertility gradient. J Ecol 61:219-224

Haridasan M (1992) Observations on soils, foliar nutrient concentrations and floristic composition of Cerrado sensu stricto and cerradão communities in central Brazil. In: Furley PA, Proctor J, Ratter JA (eds) Nature and dynamics of forest-savanna boundaries, chap 9. Chapman & Hall, London, pp 171-184

Huntley BJ (1982) Southern African savannas. In: Huntley BJ, Walker BH (eds) Ecology of tropical savannas. Springer, Berlin Heidelberg New York, pp 101-119

Isichei AO, Muoghalu JI (1992) The effects of the tree canopy cover on soil fertility in a Nigerian savanna. J Trop Ecol 8:329-338

Izaguirre-Mayoral ML, Carballo O, Flores S, de Mallorca MS, Oropeza T (1992) Quantitative analysis of the symbiotic N_2- fixation, non-structural carbohydrates and chlorophyll content in sixteen native legume species collected in different savanna sites. Symbiosis 12:293-312

Johnson NC, Zak DR, Tilman D, Pfleger FL (1991) Dynamics of vesicular-arbuscular mycorrhizae during old field succession. Oecologia 86:349-358

Johnson NC, Tilman D, Wedin D (1992) Plant and soil controls on mycorrhizal fungal communities. Ecology 73:2034-2042

Kellman M (1979) Soil enrichment by neotropical savanna trees. J Ecol 67:565-577

Lopes A S, Cox FR (1977) A survey of the fertility status of surface soils under cerrado vegetation of Brazil. J Soil Sci Am 41:742-747

Lugo AE (1988) Diversity of tropical species: questions that elude answers. Biol Int Spec Issue 19:1-37
McNaughton S (1990) Mineral nutrition and spatial concentrations of African ungulates. Nature 334:343-345
McNaughton S (1991) Dryland herbaceous perennials. In: Mooney HA, Winner WE, Pell EJ (eds) Response of plants to multiple stresses. Academic Press, San Diego, pp 307-328
Medina E (1982) Physiological ecology of neotropical savanna plants. In: Huntley BJ, Walker BH (eds) Ecology of Tropical Savannas. Ecological Studies 42. Springer Berlin Heidelberg New York, pp 308-335
Medina E (1987) Nutrient requirements, conservation and cycles in the herbaceous layer. In: Walker B (ed) Determinants of savannas. IRL Press, Oxford, pp 39-65
Medina E, Bilbao B (1991) Significance of nutrient relations and symbiosis for the competitive interaction between grasses and legumes in tropical savannas. In: Esser G, Overdieck D (eds) Modern ecology. Elsevier, Amsterdam, pp 295-319
Medina E, Huber O (1992) The role of biodiversity in the functioning of savanna ecosystems. In: Solbrig OT, van Emden HM, van Oordt PGWJ (eds) Biodiversity and global change, Monogr 8, Chap 13. Int Union Biol Sci, Paris, pp 139-158
Medina E, Silva J (1990) Savannas of northern South America: a steady state regulated by water-fire interactions on a background of low nutrient availability. J Biogeogr 17:403-413
Medina E, Mendoza A, Montes R (1982) Nutrient balance and organic matter production of the *Trachypogon* savannas of Venezuela. Trop Agric 55:243-253
Mordelet P, Abbadie L, Menaut JC (1993) Effects of tree clumps on soil characteristics in a humid savanna of West Africa (Lamto, Côte d'Ivoire). Plant Soil 153:103-111
O'Connor TG (1985) A synthesis of field experiments concerning the grass layer in the savanna regions of southern Africa. S Afr Nat Sci Progr Rep 114, Pretoria, FDR, CSIR
Oliveira-Filho ATD, Shepherd G J, Martins FR, Stubblebine WH (1989) Environmental factors affecting physiognomic and floristic variation in an area of cerrado in central Brazil. J Trop Ecol 5(4):413-431
Ratter JA, Dargie TCD (1992) An analysis of the floristic composition of 26 cerrado areas in Brazil. Edinb J Bot 49:235-250
Saif SR (1986) Vesicular-arbuscular mycorrhizae in tropical forage species as influenced by season, soil texture, fertilizers, host species and ecotypes. Angew Bot 60:125-139
Sarmiento G (1983) The savannas of tropical America. In: Bourlière F (ed) Tropical savannas. Ecosystems of the world, vol 13. Elsevier, Amsterdam, pp 246-288
Sarmiento G (1992) A conceptual model relating environmental factors and vegetation formations in the lowlands of tropical South America. In: Furley FA, Proctor J, Ratter JA (eds) Nature and dynamics of forest-savanna boundaries, Chap. 9. Chapman & Hall, London, pp 583-601
Susach F (1984) Caraterización ecológica de las sabanas de un sector de los Llanos Bajos de Venezuela. Tesis Doctoral. Univ Cent Venezuela, Fac Cienc, Caracas
Tilman D (1988) Plant strategies and the dynamics and structure of plant communities. Princeton Univ Press, Princeton
Tilman D (1990) Constraints and tradeoffs: toward a predictive theory of competition and succession. Oikos 58:3-15
Vitousek PM, Walker LR, Whiteaker LD, Mueller-Dombois D, Matson PA (1987) Biological invasion by *Myrica faya* alters ecosystem development in Hawaii. Science 238:802-804
Walker B H, Noy-Meir E (1982) Aspects of stability and resilience of savannas ecosystems. In: Huntley BJ, Walker BH (eds) Ecology of tropical savannas. Springer, Berlin Heidelberg New York, pp 577-590
Walter H (1973) Die Vegetation der Erde in ökophysiologischer Betrachtung. Band 1. Die tropischen und subtropischen Zonen. Fischer, Jena
Wedin DA, Tilman D (1990) Species effects on nitrogen cycling: a test with perennial grasses. Oecologia 84:433-441

4 Biodiversity and Water Relations in Tropical Savannas

4 Biodiversity and Water Relations in Tropical Savannas

Guillermo Sarmiento

4.1 Introduction

The analysis of savanna biodiversity and the comparison of different savanna communities is difficult due to the scanty information that exists regarding several components of the biota such as microorganisms, non vascular plants, invertebrates, and even vertebrate animals. Fortunately, enough is known of the vascular flora of several savanna communities to allow some comparisons. Unfortunately, the information on the functional groups within the vascular flora from savannas is restricted to a few study sites. This chapter reviews the savanna floristic richness based on the known vascular flora; discusses several aspects of functional diversity, especially in terms of responses to water stress, and suggests some hypothesis regarding the role of PAM and other determinants.

4.2 Biodiversity of Neotropical Savannas

4.2.1 Species Diversity

It is well known that the total number of species in any plant community depends on the size of the sampled area, as shown in Table 4.1 for the vascular flora from savanna communities from Africa and South America. From this information, the variation of species richness with the sample area can be derived, as shown in Fig. 4.1, from a minimum of 10 spp./m^2 to 100 spp./ha.

Along a soil catena in the Venezuelan western Llanos (Silva and Sarmiento 1976), the mean number of vascular plants sampled in 10 x 10 m quadrats ranged from 24.5 to 35.4 (Table 4.2). The total number of species per 1000 m^2 ranges from 53 to 85, when the ten samples from each soil type are added.

Ecological Studies, Vol. 121
Solbrig, Medina and Silva (eds) Biodiversity and Savanna Ecosystem Processes

Table 4.1. Floristic richness in some American and African savannas

Locality		
Lamto, Ivory Coast (1)	16 spp/m^2	20-35 spp/250 m^2
Nazinga, Burkina Faso (1)	8 spp/m^2	30-35 spp/1000 m^2
Calabozo, Venezuela (2)	15 spp/m^2	98 spp/520 m^2
Barinas, Venezuela (3)	25-35 spp/100 m^2	53-85 spp/1000 m^2

(1) Fournier (1991)
(2) Sarmiento and Monasterio (1969)
(3) Silva and Sarmiento (1976)

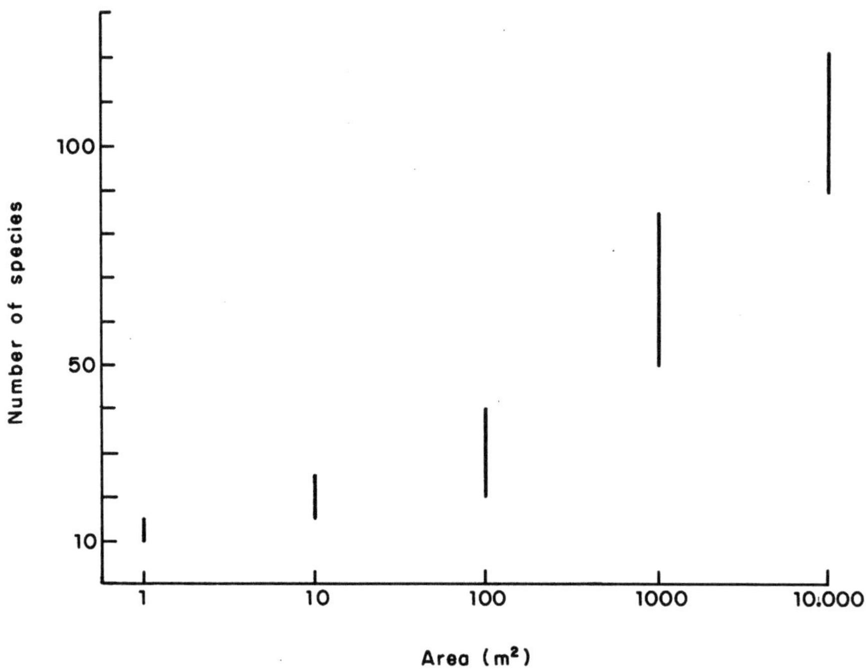

Fig. 4.1. Total number of vascular plant species according to the area sampled in different tropical savannas. The *vertical bars* indicate the range of floristic richness for each area

Table 4.2. Floristic richness in savannas of the Venezuelan llanos. (Data from Silva and Sarmiento 1976)

Savanna community	Mean no. of plant species in 10 stands of 100 m^2	Mean no. of perennial grasses in 10 stands of 100 m^2	Total no. of plant species in 1000 m^2
Palma Sola	27.4	7.3	55
Boconoito	31.6	8.6	63
Barinas	35.1	9.5	69
Camoruco	35.4	10.1	81
Garza	34.3	9.6	85
Gásperi	27.1	10.1	75
Jaboncillo	24.5	11.6	53

Table 4.3. Floristic richness of various neotropical savanna formations. (Sarmiento 1983a)

Formation	Area (km^2)	No. of. trees and shrubs	No. of sub-shrubs, half-shrubs, herbs, vines, etc.	No. of grass species	Total no. of species
Cerradoin north-western Sao Paulo (Eiten 1963)	50	45	175	14	237
Cerrado in western Minas Gerais (Goodland 1970)	15 000	~200	~330	73	~600
Whole cerrado region (Heringer et al. 1977)	2 000 000	429 (774)	181	108	718 (1063)[a]
Rio Branco savannas (Rodriguez 1971)	40000	40	87	9	136
Rupununi savannas (Goodland 1966)	12 000	~50	291	90	431
Northern Surinam savannas (Van Donselaar 1965)	~3 000	15	213	44	272 (445)[b]
Central Venezuelan llanos (Aristeguieta 1966)	3	69 (16)	175	44	288
Venezuelan llanos (Ramia 1974)	250 000	43	312	200	555
Colombian llanos (FAO 1966)	150 000	44	174	88	306

[a] Total flora including other plant formations.
[b] Total flora including bushes.
[c] Number of savanna trees excluding groves.

Considering the total vascular flora, the Brazilian cerrados are the richest neotropical savannas (Table 4.3) with more than 400 woody species and about 300 herbaceous species (Eiten 1963; Toledo Rizzini 1963; Goodland 1970; Heringer et al. 1977). Goodland (1966) reports 431 species for the Rupununi savannas in Guyana, and Van Donselaar (1965) reports 272 species for northern Surinam savannas.

In Calabozo, Venezuela, the total number of species is 288 in an area of 390 ha (Aristeguieta 1966). This figure includes the open savanna-grassland and the small isolated forest patches. The number of species in the open savanna is 175.

In West Africa, Fournier (1991) reports 130 species for the vascular flora from the Experimental Station in Lamto, Ivory Coast, and 200 species for somewhat drier savannas of Nazinga, Bourkina Faso. Floristic richness in these moist savannas seems to increase as the climate becomes drier, and this occurs in both woody and herbaceous species. In the latter, the increase is mainly due to a greater number of annual species. To the North, in the much drier Sahelian savannas from Fété Olé in Senegal, Bourlière (1978) listed 103 herbs and 22 woody species for a total of 125 vascular species.

Considering the species diversity in the grass family, the most important family of vascular plants in savannas, Sarmiento (1983b) showed that in forty 100 m^2 stands from Venezuelan savannas the number of perennial species ranged from 3 to 12, and the mean value was 7.2 species per stand.

Concerning evenness, medium values seem to be the rule in West African savannas (Fournier 1991) as well as in western Venezuelan savannas (Fariñas, pers. comm.).

The figures presented above suggest that although tropical savannas are certainly poorer in species than the humid tropical forest, they are relatively rich plant formations in the upper part of the range for terrestrial plant communities.

4.2.2 Ecological Diversity

Besides specific diversity, it would be convenient to consider other aspects of diversity such as growth forms, phenology, functional groups, and plant strategies. Tropical savannas seem to maintain rather higher levels of structural, morphological, and functional diversity. This is due to the varying proportion of woody and nonwoody elements, which results in a wide range of structural types from almost closed forest to pure grasslands. Morphological diversity is produced through the combination of the three equally important woody, half-woody, and herbaceous growth-forms. Functional diversity exists because of the coexistence of species of various types such as evergreen or deciduous, shallow or deep-rooted, perennial or annuals, and C_3 or C_4. Moreover, vegetative and reproductive phenologies show a wide range of annual patterns; some species grow only during the wet season,

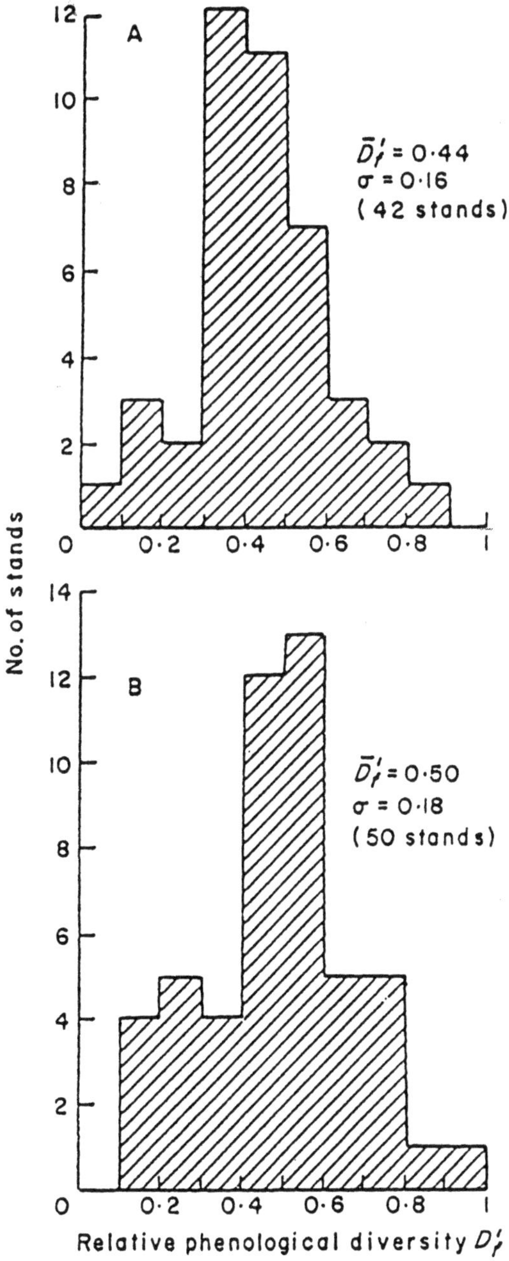

Fig. 4.2. Relative phenological diversity, per savanna stand in *A* a Venezuelan sample of 42 stands; *B* a sample of 50 stands in the western Venezuelan western llanos. $Df = \frac{Df - Df\,\mathrm{min}}{Df\,\mathrm{max} - Df\,\mathrm{min}}$ $(Df = 1/\Sigma i\, pi^2)$. (Sarmiento 1983b)

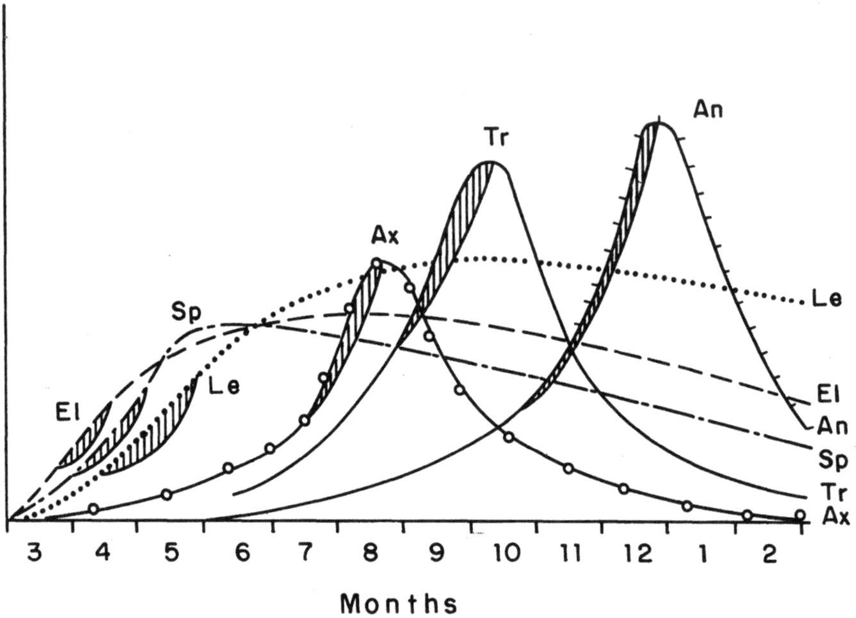

Fig. 4.3. The annual cycle of the green biomass of the six dominant grasses in the seasonal savanna in the Venezuelan llanos normally burnt in March. *Vertical hatching* indicates the flowering periods. Note the displacement of flowering and the nonoverlapping of periods of maximum groth in each species, suggesting a temporal division of the niche in the grass layer. **El** = *Elyonurus adustus;* **Sp** = *Sporobolus cubensis;* **Le** = *Leptocoryphium lanatum;* **Ax** = *Axonopus canescens;* **Tr** = *Trachypogon vestitus;* **An** = *Andropogon semiberbis.* (Sarmiento and Monasterio 1983)

others only during the dry season, whereas others grow continuously throughout the year (Monasterio and Sarmiento 1976; Lamotte 1978; Sarmiento and Monasterio 1983; Fournier 1991). Functional diversity is present within each life-form, as has been shown for evergreen trees and perennial tussock grasses (Sarmiento et al. 1985; Sarmiento 1992).

As shown in Figs. 4.2 and 4.3, an even mixture of contrasting phenological patterns characterizes each community. Furthermore, the species also differ in demographic traits such as fecundity, seed bank dynamics, and germination (Silva and Ataroff 1985).

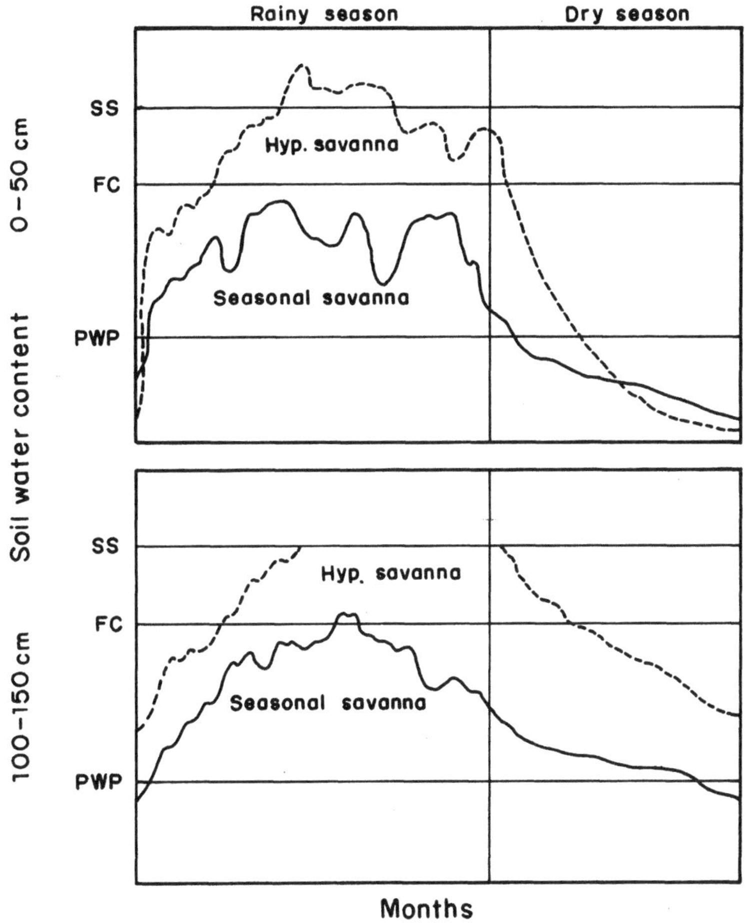

Fig. 4.4. Annual patterns of soil water content at 0-50 and at 100-150 cm in two contrasting types of tropical savannas. *SS* Soil saturation; *FC* field capacity; *PWP* permanent wilting point

4.3 Patterns of Water Availability

Plant-available moisture (PAM) is one of the crucial ecological limitations for the growth of savanna plants. PAM varies both spatially, according to depth, and temporally as a result of seasonal rainfall.

In seasonal savannas, soil water potentials are above the permanent wilting point during the rainy season, and consequently moisture is available for any plant species (Fig. 4.4). After the rainy season ends, the topsoil's water potential drops below -1.5 MPa and it remains at values as low as -3.0 to -4.0 MPa during the dry season (San Jose and Medina 1975; Sarmiento et al. 1985; Goldstein and Sarmiento 1987; Sarmiento and Acevedo 1991). Deeper, but still within the reach of woody species, soil water potentials drop slowly, remaining higher than in the topsoil for a longer portion of the dry season (Fig. 4.4). Thus, the uppermost layers show sharper contrasts in Plant Available Moisture (PAM) from one season to the next, while deeper layers are much less seasonal. Grasses and trees certainly compete for topsoil moisture, but the deeper water is available only to deep-rooted species such as the evergreen trees.

In hyperseasonal savannas, a PAM-limited dry season alternates with a rainy season during which soil is saturated and even waterlogged. Consequently, plants experience anoxic soil conditions derived from excess of water, that induce not only functional stresses but also a high mortality of fine roots (Joly 1991).

4.4 Water Loss and the Diversity of Plant Responses

The transpiratory behavior of several species of trees and perennial grasses from various savanna regions is now well documented (Bate et al. 1982; Medina 1982; Goldstein et al. 1986, 1990; Sarmiento and Acevedo 1991). Deciduous tree species respond to water stress by shedding their foliage during the dry season. Evergreen trees, on the other hand, seem to have enough available water to cover the evaporative demands of the atmosphere throughout the year. Four evergreen trees, very characteristic of neotropical savannas – *Curatella americana*, *Byrsonima crassifolia*, *Bowdichia virgilioides* and *Casearia sylvestris* – show seasonal courses of transpiration rates more related to air VPD (Vapor Pressure Deficit) than to the alternating of dry and moist seasons (Fig. 4.5; Goldstein and Sarmiento 1987). Perennial grasses evidence a clear seasonal behavior, since they decrease their transpiration rates sharply during the rainless season (Goldstein and Sarmiento 1987).

There are important differences between the species' responses within each of these functional groups. Evergreen tree species differ in their daily patterns of stomatal conductance and transpiratory fluxes, and some of them severely restrict water losses during the midday hours (Fig. 4.6). Of the four species mentioned above, *Bowdichia virgilioides* is the most drought-avoiding species, since it shows the highest turgor-loss point, the lowest resistance to water flow, and the severest stomatal control. *Casearia sylvestris*, in contrast, is the most drought-tolerant species because it shows the lowest turgor-loss point and the most negative leaf water potential. *Curatella americana* and *Byrsonima crassifolia* seem to have intermediate drought-resistant strategies.

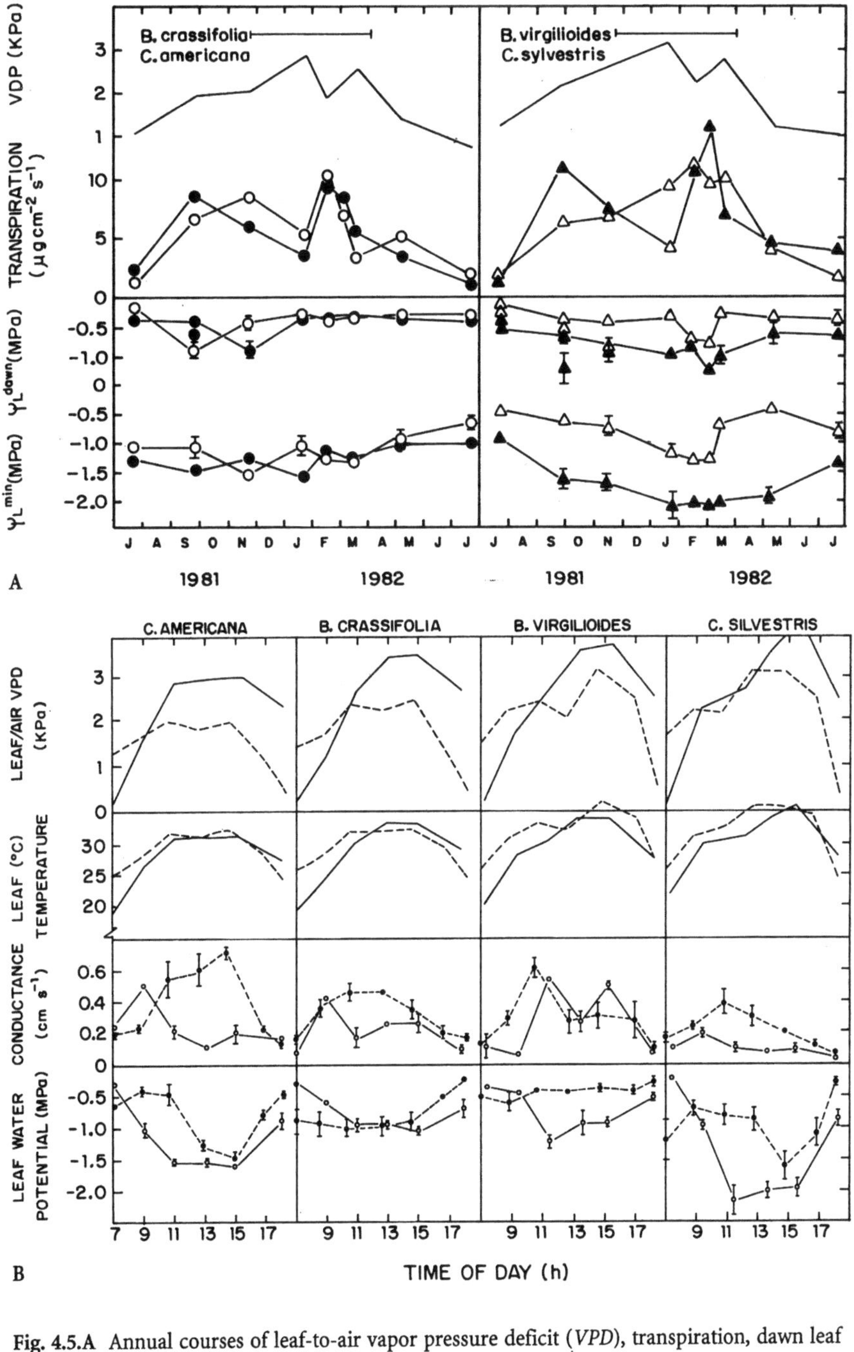

Fig. 4.5.A Annual courses of leaf-to-air vapor pressure deficit (*VPD*), transpiration, dawn leaf water potential, (ψ_L^{dawn})and minimum leaf water potential (ψ_L^{min}) in four woody species of the Venezuelan savannas. Vertical bars represent 1 ± standard error; absence of bars indicates that the standard error was smaller than the symbol. Upper segments indicate length of the dry season (Goldstein et al. 1986). **B** Daily courses of leaf-to-air vapor pressure deficit (VPD), leaf temperature, stomatal conductance, and leaf water potential for four woody species of the Venezuelan savannas. Dashed Lines correspond to one wet season day, the continuous line to one dry season day. (Goldstein et al. 1986)

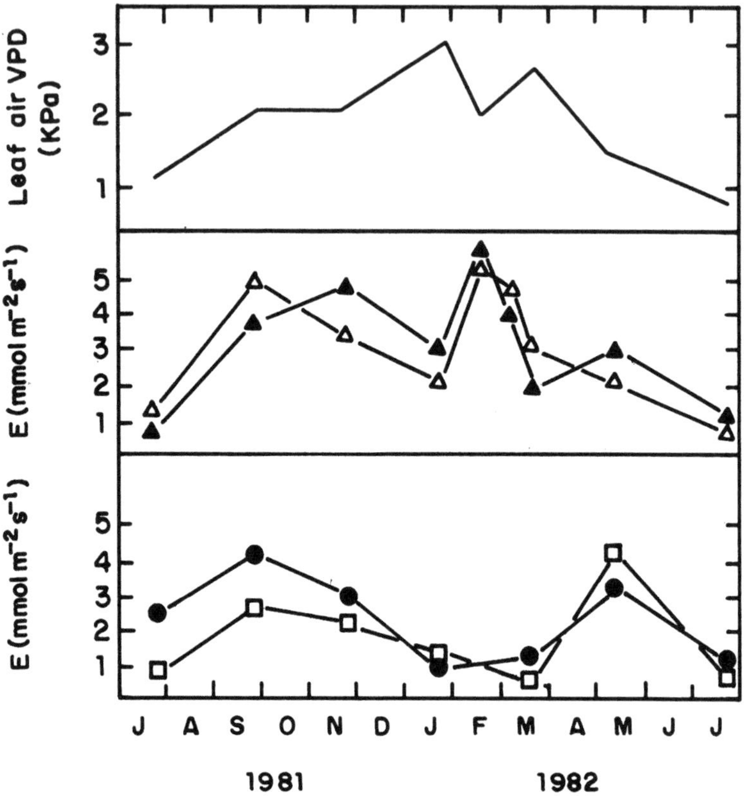

Fig.4.6. Seasonal courses of leaf-to-air vapor pressure deficit *(VDP),* and transpiration rates *(E)* for *Curatella americana* (Δ) *Byrsonima crassifolia* (▲), *Sporobolus cubensis* (□), and *Andropogon semiberbis* (●). The first two species are evergreen trees and the last two are perennial C4 grasses. (Goldstein et al. 1986 and unpubl. data)

Grasses also differ from each other in the minimum leaf water potential and in the actual transpiratory fluxes that they attain in the dry-season days (Goldstein and Sarmiento 1987). Concerning minimum leaf-water potential, *Sporobolus cubensis* shows a gradual decrease in values throughout the dry season, while *Trachypogon vestitus* consistently maintains lower values, and experiences a marked drop that starts in the second half of the wet season (Fig. 4.7). *Sporobolus cubensis* also exhibits a lower transpiration rate than *Andropogon semiberbis* during most of the year (Fig. 4.7). Grass species may also show contrasting behavior under water stress, when they attain low stomatal conductance. Some species, such as *Trachypogon vestitus* and *Hyparrenia rufa,* maintain assimilation rates at low conductance relatively higher than others like *Leptocoryphium lanatum* (Goldstein and Sarmiento 1987).

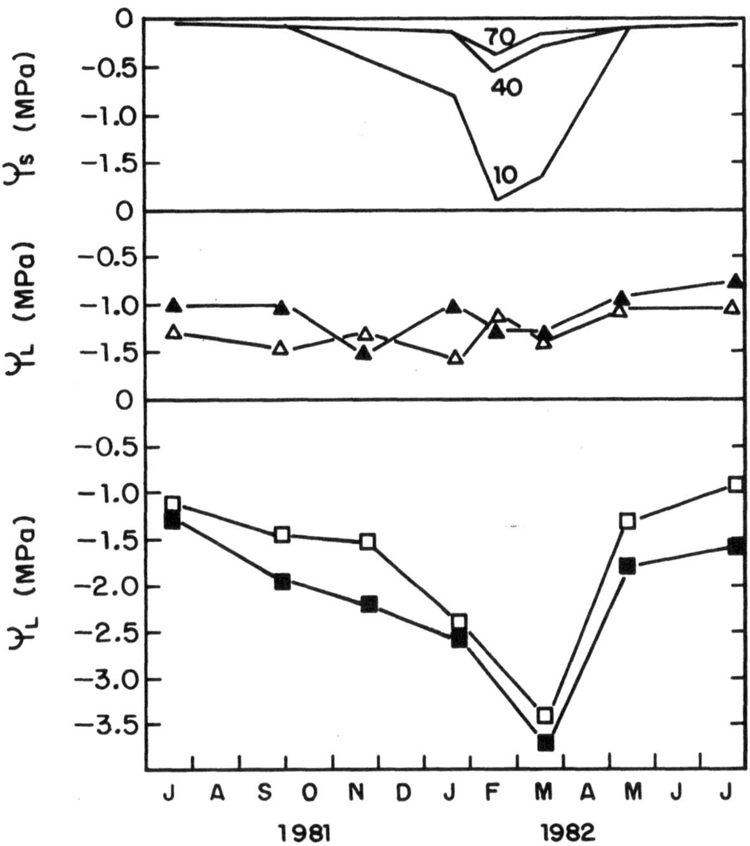

Fig.4.7. Seasonal courses of soil water potential (Ψ soil) *at 10, 40, and 70 cm depth, and minimum leaf water potential* (ΨL) for *Curatella americana* (Δ), *Byrsonima crassifolia* (▲), *Sporobolus cubensis* (□), and *Trachypogon vestitus* (■). The first two species are evergreen trees and the last two are perennial C_4 grasses. (Goldstein et al. 1986 and unpubl. data)

The pattern emerging from the diversity in grass responses is related to phenological diversity. In fact, precocious species such as *S. cubensis* and *L. lanatum*, that start regrowth and bloom immediately after the onset of the rains or even after a fire at the end of the dry season, appear to be less drought-tolerant than intermediate and late species such as *T. vestitus*, *A. semiberbis*, and *H. rufa*. Early-growing species have higher water use efficiency than lategrowers; consequently, they grow best when water is available, and they are less tolerant to adverse water conditions than late-growing species.

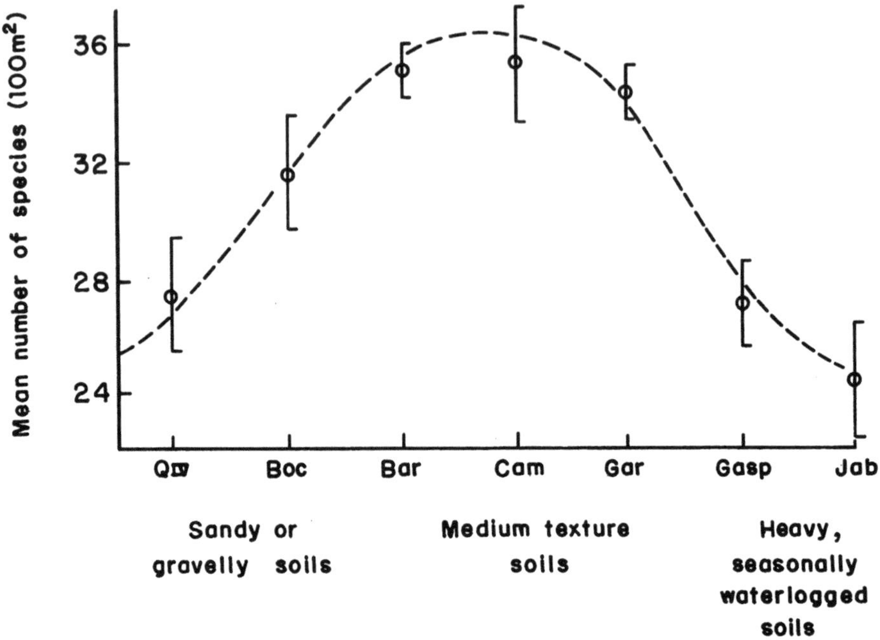

Fig. 4.8. Mean number of species of vascular plants in ten samples (100 m^2) from seven savanna communities in the Venezuelan western llanos, ordered along a soil moisture gradient. (Data from Silva and Sarmiento 1976)

4.5 Plant-Available Moisure and the Specific and Functional Diversity of Savannas

Differences in history and evolution normally tend to obscure the relationships between diversity and function in plant communities. Therefore, to avoid this, comparisons are made between communities within the same or in closely related geographical areas. Here, we will compare savanna communities from the western Venezuelan Llanos based on data from Silva and Sarmiento (1976).

Diversity, as total plant species richness, was measured along a topographic gradient (Fig. 4.8). Soil moisture and PAM vary along the gradient from well-drained soils with a long seasonal drought at one extreme, to poorly drained soils with both a long drought and a season of flooding at the other; the central part of the gradient seems to present the most favorable soil water conditions, since it is free from waterlogging and has a long PAM season. Richness is maximum under these conditions. Other measurements of community diversity, such as richness and evenness of perennial grass species, show the same pattern of variation along the gradient. In conclusion, maximum diversity corresponds to lowest water stress

Other soil properties also change along the gradient. The dystrophic condition of the soil ameliorates from the well-drained to the poorly drained conditions. Thus, in this case, diversity responds to PAM but not PAN (Plant Available Nutrients).

Similar results were found in a more extensive study on a wide range of Venezuelan seasonal savannas (Sarmiento 1983b). Floristic diversity, measured as perennial-grass species richness, also peaks toward the more mesic conditions and decreases towards both extremes: dry sandy soils and wet, seasonally saturated soils.

Phenological diversity in the perennial grass component presents a somewhat more confused picture. Richness and evenness are higher at intermediate, neither too dry nor too wet, climatic and edaphic conditions. However, there is a tendency for each phenological group to become dominant under particular ecological conditions; precocious species on deep soils and a relatively long rainy season; early species on deep soils but a relatively short rainy season; intermediate grasses on shallow soil with a short rainy season, and late species on shallow soil with a long rainy season. The picture is far from clear, and other factors like fire and nutrients may be involved.

4.6 Conclusions

There is a hierarchy of determinants, some of them correlated to each other, that affect species composition and diversity of savanna communities. At the top of the hierarchy is the annual regime of PAM, but other determinants such as PAN, fire, and herbivory must also be considered (Solbrig 1991).

Short-term changes and period pulsation of these determinants may alter species composition and diversity. However, the functional diversity of savannas allows for only minor floristic changes. "Redundant" species within each functional group are not entirely equivalent, and hence they have different tolerances to environmental pulsation. Therefore, changing conditions may result in a certain degree of floristic replacement, without major changes in the functioning of the system. This means that savannas should be more stable in functional than in floristic terms. Only major changes in the top levels of the hierarchy of environmental determinants conceivably lead to functional changes and eventually to the complete replacement of the savanna ecosystem.

References

Aristeguieta L (1966) Flórula de la Estación Biológica de Los Llanos. Bol Soc Venez Cienc Nat 110:228-307

Bate GC, Furniss PR, Pendle BG (1982) Water relations of Southern African savannas. In: Huntley BJ, Walker BH (eds) Ecology of tropical savannas. Springer, Berlin Heidelberg New York pp 336-358

Bourlière F (1978) La savane sahélienne de Fété-Olé, Sénégal. In: Lamotte M Bourlière F (eds) Problémes d'ecologie, ecosystèmes terrestres. Masson, Paris, pp 187-229

Eiten G (1963) Habitat flora of Fazenda Campiniha, Sao Paulo, Brazil. In: Ferri MG (ed) Simpósio Sobre O. Cerrado. Univ Sao Paulo, Sao Paulo, pp 179-231

FAO (1966) Reconociemiento Edafológico de los Llanos Orientales, Colombia. Tomo III. La Vegetación Natural y la Ganadería en los Llanos Orientales. Fao, Roma

Fournier A (1991) Phénologie, croissance et production végétales dans quelques savanes d'Afrique de l'Ouest. Variation Selon un Gradient Climatique. Orstom, Paris

Goldstein G, Sarmiento G (1987) Water. Water relations of trees and grasses and their consequences for the structure of savanna vegetation. In: Walker BH (ed) Determinants of tropical savannas. IUBS Monogr Ser 3, IRL Press, Oxford, pp 13-38

Goldstein G, Sarmiento G, Meinzer R (1986) Patrones diarios y estacionales en las relaciones hidricas de árboles siempreverdes de la sabana tropical. Acta Oecol, Oecol Plant 7:107-119

Goldstein G, Rada F, Canales J, Azocar A (1990) Relaciones hidricas e intercambio de gases en especies de sabanas americanas. In: Sarmiento G (ed) Las Sabanas Americanas. Aspectos de su biogeografia, ecologia y utilización. Fondo Editorial Acta Cient Venez, Caracas, pp 219-242

Goodland R (1966) South American savannas. Comparative studies Llanos and Guyana. McGill Univ Savanna Res Ser 5:1-52

Goodland R (1970) Plants of the cerrado vegetation of Brazil. Phytologia 20:57-78

Heringer EP, Barroso GM, Rizzo JA, Rizzini CT (1977) A flora do cerrado. In: Ferri MG (ed) IV Simpósio sobre o Cerrado. Univ Sao Paulo, Sao Paulo, pp 211-232

Joly CA (1991) Adaptações de plantas de savanas e florestas neotropicais a inundação. In: Sarmiento G (ed) Las Sabanas Americanas. Aspectos de su biogeografía, ecologia y utilización. Fondo Editorial Acta Cient Venez, Caracas, pp 243-257

Lamotte M (1978) La savane préforestière de Lamto, Cote d'Ivoire. In: Bourlière F, Lamotte M (eds) Problémes d'ecologie, ecosystèmes terrestres. Masson, Paris, pp 231-311

Medina E (1982) Physiological ecology of neotropical savanna plants. In: Huntley BJ, Walker BH (eds) Ecology of tropical savannas. Springer Berlin Heidelberg NewYork, pp 308-335

Monasterio M, Sarmiento G (1976) Phenological strategies in species of seasonal savanna and semi-deciduous forest in the Venezuelan llanos. J Biogeogr 3:325-355

Ramia M (1974) Plantas de las Sabanas Llaneras. Monte Avila, Caracas

Rodriguez WA (1971) Plantas dos campos do Rio Branco (Território de Roraima). In: Ferri MG (ed) III Simpósio sôbre o cerrado. Univ Sao Paulo, Sao Paulo, pp 180-193

San José JJ, Medina E (1975) Effect of fire on organic matter production and water balance in a tropical savanna. In: Golley FB, Medina E (eds) Tropical ecological systems. Springer, Berlin Heidelberg New York, pp 251-264

Sarmiento G (1983a) The savannas of tropical America. In: Bourlière F (ed) Tropical savannas. Elsevier, Amsterdam, pp 245-288

Sarmiento G (1983b) Patterns of specific and phenological diversity in the grass community of the Venezuelan tropical savannas. J Biogeogr 10:373-391

Sarmiento G (1992) Adaptive strategies of perennial grasses in South American savannas. J Vegetat Sci 3:325-336

Sarmiento G, Acevedo D (1991) Dinámica del agua en el suelo, evaporación y transpiración en una pastura y un cultivo de maiz sobre un alfisol en los Llanos Occidentales de Venezuela. Ecotrópicos 4:27-42

Sarmiento G, Monasterio M (1969) Studies on the savanna vegetation of the Venezuelan Llanos. J Ecol 57:579-598
Sarmiento G, Monasterio M (1983) Life forms and phenology. In: Bourlière F (ed) Tropical savannas. Elsevier, Amsterdam, pp 79-108
Silva J, Ataroff M (1985) Phenology, seed crop and germination of coexisting grass species from a tropical savanna in western Venezuela. Acta Oecol Oecol Plant 6:41-51
Sarmiento G, Goldstein G, Meinzer F (1985) Adaptive strategies of woody species in neotropical savannas. Biol Rev 60:315-355
Silva J, Sarmiento G (1976) La composition de las sabanas en Barinas en relación con las unidades edáficas. Acta Cient Venez 27:68-78
Solbrig OT (1991) Savanna modelling for global change. Biol Int Spec Issue 24:47, IUBS, Paris
Toledo Rizzini C (1963) A flora do cerrado. In: Ferri MG (ed) Simpósio Sobre o Cerrado. Univ Sao Paulo, Sao Paulo, pp 125-177
Van Donselaar J (1965) An ecological and phytogeographic study of northern Surinam savannas. Wentia 14:1-163

5 Ecophysiological Aspects of the Invasion by African Grasses and Their Impact on Biodiversity and Function of Neotropical Savannas

5 Ecophysiological Aspects of the Invasion by African Grasses and Their Impact on Biodiversity and Function of Neotropical Savannas

Zdravko Baruch

5.1 Introduction

Biological invasions expand the geographical range of plants and animals. They proceed in relation to the organism's physiological tolerance, migratory ability, and biogeographical barriers. During Earth's history, these natural processes have been responsible for the assembly of communities through coadaptation and coexistence of species of distinct origin in a common environment. With human mediation, biological invasions have been gradually accelerated. In neotropical savannas, inadvertent or deliberate introduction of African grasses since colonial times was followed by spontaneous spread of the invaders. More recent introductions in this century, for the purpose of pasture improvement, have spread out rapidly and partially displaced native grasses. These new forage grasses were favored by the stockowners, due to their better persistence under grazing and higher nutrient value compared to the indigenous grasses (Parsons 1972).

Once established, the range of alien grasses expanded swiftly (Vareschi 1970; Baruch et al. 1989; San José and Fariñas 1991) probably aided by the recurrent opening of the native savanna by fire (Orians 1986), by their preadaptation to environments similar to those in their native African savannas, by their high dispersion potential (due to low seed weight and high germinability), and by their grazing tolerance and immunity to pathogens in their new habitat (Baker 1978; Baruch et al. 1989; Simoes and

Ecological Studies, Vol. 121
Solbrig, Medina and Silva (eds) Biodiversity and Savanna Ecosystem Processes

Table 5.1. Species richness of herbaceous species in Venezuelan savanna communities dominated by native and by alien grasses

	Total no. of herbaceous plants	No. of spp /m^2
1. Highland savannas, coastal mountains (1000 - 1300 m asl)		
a) Dominated by native grasses *T. plumosus* and *Axonopus pulcher*	104 - 114 (1, 2)	5 - 14 (1, 2)
b) Dominated by the alien grass *M. minutiflora*	-----------	2 (2)
2. Lowland nonflooded savannas (50 - 200 m asl)		
a) Dominated by native grasses *Trachypogon* spp. and *Axonopus canescens*	70 - 107 (3, 4, 5)	9 - 15 (3, 4)
b) Dominated by the alien grass. *H. rufa*	"few" - 20[a] (6, 7, 8, 9)	2 - "few" (6, 7, 8)

(1) Baruch (1986); (2) Vareschi (1970); (3) Sarmiento and Monasterio (1969); (4) Velásquez (1965); (5) Aristeguieta (1966); (6) Pieters (1993); (7) Cruces (1977); (8) San José and Fariñas (1991); (9) Bulla, this Volume.

[a] Includes other species in nearby areas.

Baruch 1991; Klink 1994, 1995). In Venezuela, four African grasses are the most prominent invaders: *Melinis minutiflora* Beauv. in derived or secondary savannas above 600 m asl.; *Hyparrhenia rufa* (Nees) Stapf. in lowland savannas with poor soils and marked dry seasons; *Panicum maximum* Jacq. in humid and relatively fertile areas, and *Brachiaria mutica* (Forsk.) Stapf. in periodically flooded savannas. The successful encroachment of the alien grasses generally took place only in the wetter and/or more fertile habitats of the savanna. In more stressful sites, the indigenous community persists and it is generally dominated by species of the graminoid genera *Trachypogon, Axonopus,* and *Bulbostylis.* (Baruch et al. 1989; Pieters 1993).

In Venezuela, savannas are located mainly between the Andean and Coastal mountains, and the Guiana plateau. Their physiognomy generally combines an open tree stratum immersed in a continuous herbaceous matrix. The soils are highly leached oxisols and ultisols, very low in exchangeable bases and high in aluminum. The climate is warm throughout the year and rainfall is highly seasonal with 800 - 2000 mm/year (Sarmiento 1984; Medina and Silva 1990). The following discussion is based on the studies of the invasions of *M. minutiflora* in some relatively fertile highland savannas, and of *H. rufa* in the less fertile lowland Llanos of Venezuela. Both sites were originally dominated by several species of native tussock grasses belonging to the genera *Trachypogon* and *Axonopus.*

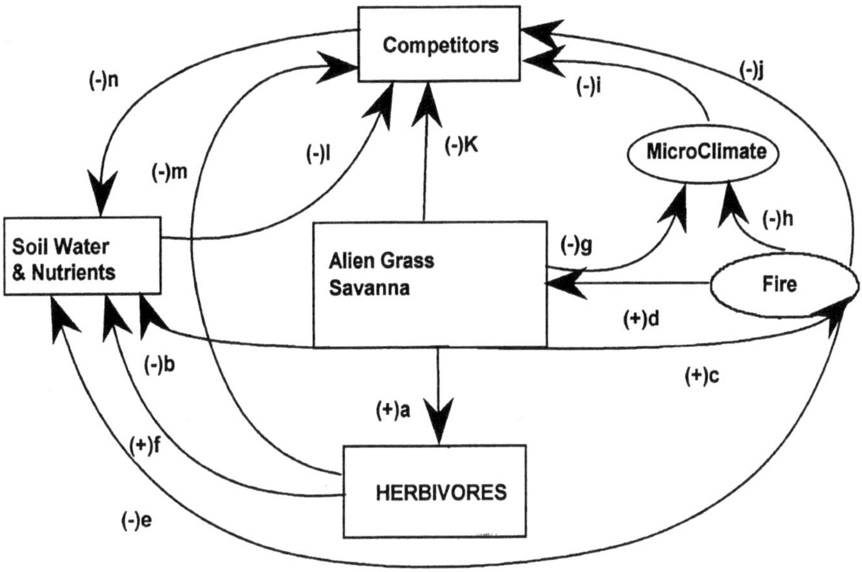

Fig 5.1. Schematic model of the hypothetical effects of replacing native savannas with communities dominated by alien grasses. *Plus* and *minus symbols* indicate positive or negative effect. *a* Alien grasses increase primary production (PP) increasing forage supply for herbivores; *b* increased PP demands more nutrients and water from the soils reducing their availability for other plants; *c* increased PP provides more fuel for frequent and more intense fires; *d* fires open further savannas for invasion by alien grasses; *e* fires increase volatilization and losss of mineral nutrients; *f* increased consumption by herbivores accelerates nutrient cycling; *g* the closed canopy of alien grasses affects the microclimate; *h* increased fire frequency and intensity also affect the microclimate; *i* microclimate alteration impedes establishment and survival of native competitors of alien grasses; *j* fires displace fire-intolerant competitors of alien grasses; *k* alien grasses displace competitors through competition by space and by allelopathy; *l* alien grasses outcompete native plants for water and nutrient use; *m* herbivory damage more native plants than alien grasses; *n* displaced native legumes reduce symbiotic nitrogen fixation

Alien grasses have displaced the indigenous savanna species over large areas developing almost monospecific stands with 100 % cover and drastically reduced species diversity (Table 5.1). In addition to changes in species number and composition, the physiognomy of the grassland was also changed: the stoloniferous *M. minutiflora* builds stands about 1 m tall formed by a compact cushion of culms and live and dead leaves, whereas *H. rufa*, a bunch grass, forms dense closed stands nearly 3 m tall. The medium-tallgrass native savanna is thereby converted into a tallgrass savanna. The reduced diversity and simplified structure of grasslands dominated by alien grasses probably alter the savanna's biogeochemical cycles by: (1) shifting the rates of resource supply (water, light, and mineral nutrients) (2) modifying the trophic structure of the system (changes in primary productivity and food chain links), and (3) increasing the disturbance rate (the frequency and intensity of fire and herbivory). This chapter discusses some ecophysiological differences between indigenous and introduced grasses accounting for the success of the latter in neotropical savannas, and considers the possible effects that the concurrent decrease in specific and structural diversity might have on the function of the savanna ecosystem as shown in a schematic model (Fig. 5.1.).

5.2 Photosynthesis, Growth, and Primary Production

All native and introduced grasses considered here are C_4 plants which have similar light saturation curves, quantum yield (0.050 - 0.058 mol CO_2/ mol quanta) and optimum temperature for photosynthesis (Fig. 5.2a; Baruch 1989). However, the African grasses tend to have higher growth rates both under field (Baruch et al. 1989; Klink 1994) and controlled conditions (Baruch 1986), possibly resulting from their higher photosynthetic capacity and to the larger relative allocation of assimilates to leaves with higher specific leaf area compared to native grasses (Baruch et al. 1985; 1989) (Table 5.2 and Fig. 5.2a). In contrast, root/shoot ratios are higher in native grasses, most of their carbon reserves being stored below ground (Baruch et al. 1989; Simoes and Baruch 1991). The high growth rates of the introduced grasses are typical of invader plants (Bazzaz 1986) and are consistent with their ability to establish rapidly, compete successfully for resources, and displace native species from moist, fertile sites. Conversely, the slower growth rate of the native grasses is consistent with their capacity to endure invasion and persist in poorer sites such as the infertile patches of shallow lithoplinthic soils of the lowland nonflooded savannas of Venezuela (Pieters 1993).

As a consequence of their higher growth rates, the biomass accumulation in the almost pure stands of introduced grasses is significantly higher than that of the native grasslands (Table 5.2; Cruces 1977). Although herbivores generally consume a small fraction of the available plant biomass in neotropical savannas, the higher biomass of the communities dominated by alien grasses offers additional resources for herbivores and their predators, possibly altering their respective population dynamics (Fig. 5.1a). In particular, the activity of termites and ants in decomposition and mineralization of plant dead biomass may also be enhanced (Fig. 5.1f) (Medina 1993). However, the decrease in species richness and structural heterogeneity accompanying the displacement of the indigenous savanna would

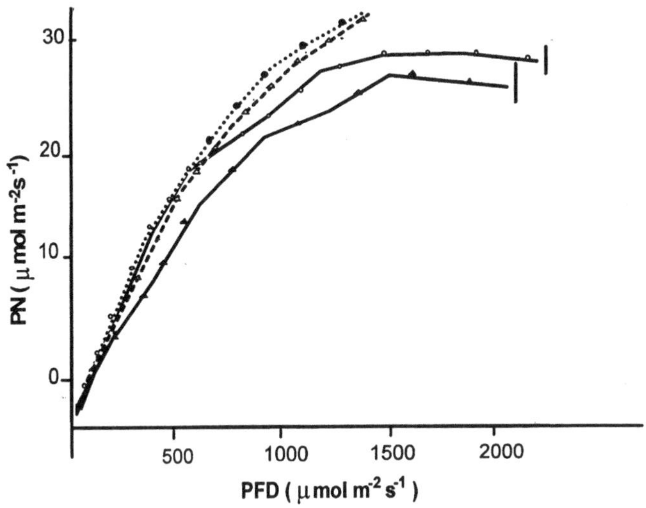

a

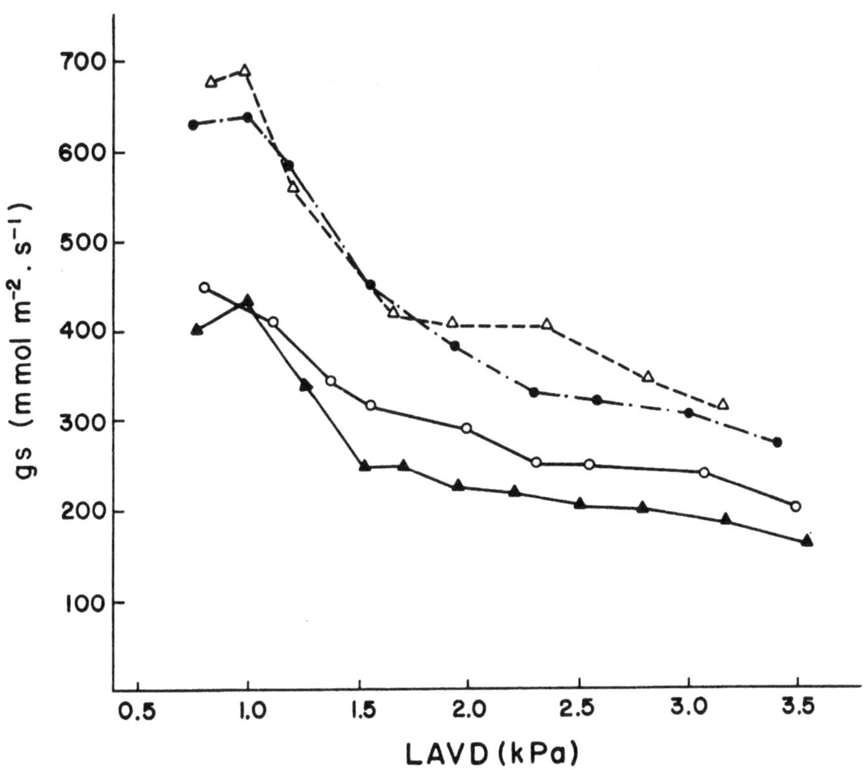

b

Fig. 5.2. a.b. Legends see Page 84

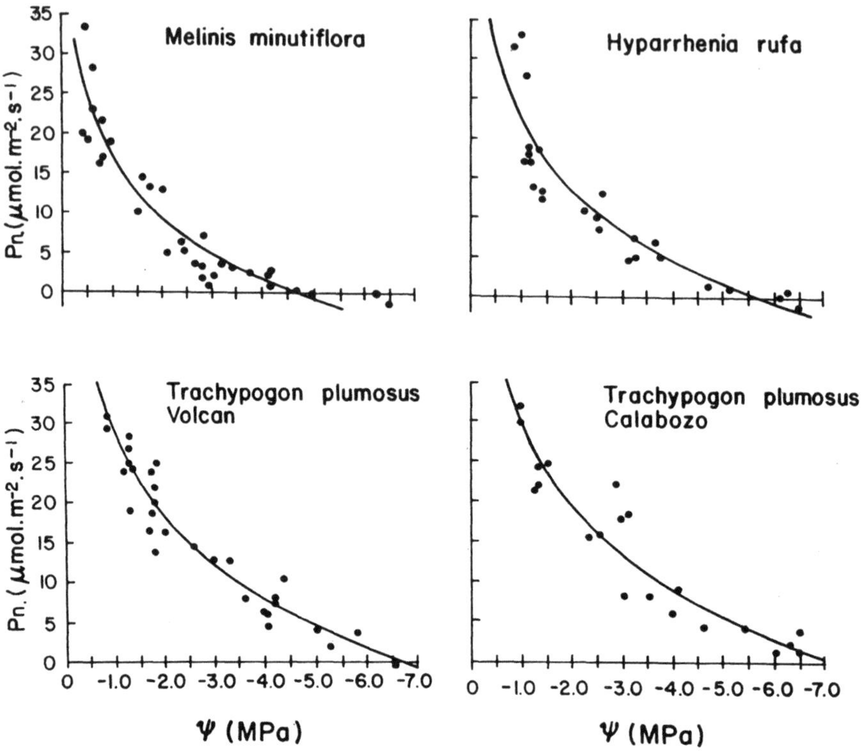

Figure 5.2. *a* Relationship between net photosynthetic rate (*Pn*) and photon flux density (*PFD*) in two introduced African grasses (*broken lines; open triangles* = *H. rufa* and *solid circles* = *M. minutiflora*) and the native *T. plumosus* (*solid lines; solid triangles* = lowland population of Calabozo and *open circles* = highland population from El Volcán). Bars are standard errors for five to ten leaves for species or population; *b* influence of leaf-air vapor pressure difference (*LAVD*) on stomatal conductance (*gs*). Species and symbols as in *a*; *c* influence of leaf water potential (ψ) on net photosynthesis rate (*Pn*) in native and alien grasses. The relationshs were obtained over a slow soil-drying cycle that lasted 16-40 days and the *lines* represent the best fit (P<0.01 %). *Arrows* indicate when Pn approaches zero. (Baruch et al 1985)

lead to a loss of diversity and reduced persistence of the native animal populations that depend on native plants for nourishment and refuge. In addition, the large standing dead biomass left by alien grass communities at the end of the dry season facilitates combustion and increases fire intensity (Fig. 5.1c), altering the microclimate (Fig. 5.1h) and resulting in larger losses of nitrogen and sulfur through volatilization compared to that of the native grasslands (Fig. 5.1e) (Medina 1993). Analogous to the Hawaiian grass / fire cycle, it is likely that the processes of invasion and persistence of African grasses in neotropical savannas are sustained by a fire and invasion cycle. Native communities opened by fire are invaded by alien grasses, which in turn promote more frequent and intense fires, reinforcing the process of colonization (Fig. 5.1d) and eliminating native competitors (Fig. 5.1j) (D'Antonio and Vitousek 1992).

Table 5.2. Growth rate, above ground biomass and biomass allocated to leaves in native and alien grasses from highland savannas (HS) and lowland savannas (LS) in Venezuela

	T. plumosus (HS)	*M. minutiflora* (HS)	*T. plumosus* (LS)	*H. rufa* (LS)
Maximum leaf elongation rate (cm/day)	2.7 ± 0.4 (1) (*)	3.7 ± 0.5 (1) (*)	1.7 ± 0.3 (2) (*)	3.0 ± 0.7 (2) (*)
Maximum leaf area ratio (cm^2/ g dry wt.)	18.2 ± 3.0 (1) (+)	56.2 ± 8.4 (1) (+)	30.6 ± 5.7 (2) (*)	144.2± 27.3(2) (*)
Maximum leaf biomass (%)	17.2 ± 2.7 (3) (+)	31.0 ± 8.3 (3) (+)	20.1 ± 4.5 (2) (*)	50.1 ± 3.1 (2) (*)
Peak above ground biomass (g / m^2)	Not available	Not available	411-731 (4, 5) (+)	500-1400 (6,7) (+)

(1) Baruch (1989); (2) Simoes and Baruch (1991); (3) Baruch et al. (1989); (4) Rodriguez (1987); (5) San Jose and Medina (1976); (6) Cruces (1977); (7) Bulla, this Volume.

Asterisks (*) denote laboratory results and plus sign (+) field results

5.3 Nutrients and Nitrogen Use Efficiency

Scarcity of plant-available nutrients is one of the main environmental constraints for plant growth in seasonal savannas (Frost et al. 1986; Medina and Silva 1990). Plants generally respond by producing biomass with lower nutrient concentration. In alien grasses, nutrients are allocated preferentially to the production of large assimilating surfaces (Sarmiento 1992; Baruch and Gomez, unpubl. results). It is apparent that introduced grasses, in both American and Australian savannas, require more nutrients than native grasses in order to achieve their higher growth rates and they also seem to use nitrogen more efficiently (Table 5.3; Christie and Moorby 1985; Baruch et al. 1985; Bilbao and Medina 1990). The preferential settlement of alien grasses on the most fertile soils in the savanna (Baruch et al. 1989) confirms their higher mineral nutrient requirements. In addition, recent results show that experimental fertilization promoted the establishment of *H. rufa* on soils where it normally does not occur (Pieters 1993). In native and alien grasses from the infertile lowland savannas of Venezuela, retranslocation of assimilates and minerals from leaves to roots and rhizomes takes place during the dry season (Table 5.3; Baruch and Gomez, unpubl. results). This accumulation of reserves in underground organs during the dry season probably contributes to their preservation from fire volatilization,

Table 5.3. Nitrogen use efficiency (NUE) and leaf nitrogen and phosphorus concentration during the rainy and dry seasons in native and alien grasses.

	T. plumosus (HS)	*M. minutiflora* (HS)	*T. plumosus* (LS)	*H. rufa* (LS)
NUE (*)				
(Mmol CO_2 / gN s) (1)	29.6 ± 1.1	38.0 ±1.2	31.8 ± 2.6	40.3 ± 1.4
Nitrogen (+) (mg g/1) (2)				
Rainy season	15.5 ± 1.2	16.4 ± 5.5	11.0 ± 3.3	8.4 ± 3.1
Dry season	13.7 ±1.6	16.9 ± 4.7	6.5 ± 0.6	2.8 ±0.9
Phosphorus (+) (mg/g) (2)				
Rainy season	1.0 ± 0.2	1.4 ± 0.2	0.5 ± 0.2	0.5 ± 0.1
Dry season	0.8 ± 0.2	1.5 ± 0.2	0.3 ± 0.1	0.2 ± 0.1

(1) Baruch et al. (1985); (2) Baruch and Gómez (unpubl. results).
(Asterisks) denotes laboratory data and (+) field data.

constituting an important conservation strategy in nutrient-poor habitats. This seasonal nutrient reallocation is less pronounced in grasses from more fertile highland savannas (Table 5.3). Retranslocation of underground reserves takes place during the initial flush of growth at the beginning of the rainy season.

Leaf nitrogen and phosphorus concentrations do not differ markedly between the alien *H. rufa* and *M. minutiflora* and the native populations of *T. plumosus* in either the highland fertile savanna or the lowland infertile savanna (Table 5.3) and the phosphorous/nitrogen ratio ranges between 0.05 and 0.09 in both groups. This low ratio indicates that phosphorus is barely sufficient for plant growth, and suggests that native and alien grasses should respond more to fertilization with phosphorus than with nitrogen (Medina 1993).

The higher biomass per unit area of the communities dominated by alien grasses indicates that the amounts and rates involved in nutrient cycling should be higher than in the native savannas. It is possible that the high nutrient requirements of alien species deplete the soil reserves (Fig. 5.1b) and there is also a potential for nutrient loss from the large standing dead biomass exposed to rainfall leaching and to volatilization by fire in the communities dominated by alien grasses (Fig. 5.1e). On the other hand, the displacement of native legumes by alien grasses should reduce the biological fixation of nitrogen (Fig. 5.1n). The combined effects of reduced fixation and increased loss of nitrogen might result in negative balances of this nutrient in grasslands dominated by alien grasses. However, comparative studies are still lacking. On the other hand, it is also possible that the large amounts of air contaminants, typical in the proximity of large cities and industrial centers, dissolved in rain water might increase soil fertility in these areas and favor the invasion of alien grasses (Berendse and Elberse 1990).

5.4 Water Relations and Water Use Efficiency

Water availability is another of the main environmental limitations for plant life in the seasonal tropical savannas (Frost et al. 1986; Medina and Silva 1990). Native and alien grasses have to compete for soil water in a process that may affect their respective carbon and nutrient capture and use. Both groups differ markedly in their strategies of water use under stressed and nonstressed conditions. Under well-watered laboratory conditions and during the rainy season in the field, alien grasses had consistently higher stomatal conductances and transpiration rates than *T. plumosus* (Fig. 5.2b), a fact partially accounting for their higher photosynthetic rates and higher water use efficiency (Baruch et al. 1985; Simoes and Baruch 1991; Baruch and Fernandez 1993). During the dry season, however, *H. rufa* responds with fast senescence and a drastic reduction of leaf area. In the more drought-tolerant *T. plumosus* and other indigenous grasses, leaf activity continues, although at a reduced level, resulting in higher water use efficiency during this period (Fig. 5.2c). Both groups of grasses show similar degrees of osmotic adjustment under drought (Baruch et al. 1985; Baruch, Ludlow and Wilson, unpubl. results) and stomatal sensitivity to soil or atmospheric water stress (Baruch et al. 1985; Baruch and Fernandez 1993). However, drought-evasion mechanisms are more conspicuous in *H. rufa*, whereas *T. plumosus* is more drought-tolerant and uses water more "conservatively". The opportunistic water use of alien grasses contrasts with that of native grasses and probably contributes to the higher competitive potential of the former in the seasonal tropical savannas (Fig. 5.1l) (Baruch et al. 1985; Baruch and Fernandez 1993). Also, the root system of *H. rufa* is deeper than that of *T. plumosus* (Baruch unpubl. results), by which the former has access to larger amounts of water (and nutrients) to support its high growth rate.

Although detailed studies are not available, it is conceivable that the establishment of communities dominated by alien grasses can alter the water balance of the area through increased evapotranspiration rates, rainfall interception, and decreasing soil water recharge and runoff in virtue of their higher leaf area index, deeper root system, and higher transpiration rates than native savanna communities (Fig. 5.1b).

5.5 Responses to Defoliation

In neotropical and Australian savannas, alien African grasses are more tolerant to defoliation than indigenous grasses (Hodgkinson et al. 1989; Simoes and Baruch 1991; Sarmiento 1992), but *M. minutiflora* seems to be the exception in the Brazilian Cerrados (Klink 1994). Although experimental defoliation reduced total plant biomass in both *H. rufa* and *T. plumosus*, only the former compensated for the leaf biomass and area lost (Fig. 5.3). This contrasting behavior has been attributed to a larger proportion of assimilates allocated to leaf and tiller production and higher leaf growth rate in *H. rufa* (Simoes and Baruch 1991). Also, prostrate growth, which reduces grazing pressure, is favored in the alien grass, whereas *T. plumosus* showed decreased tillering and increased senescence under frequent defoliation (Simoes and Baruch 1991). Both physiological and architectural responses to defoliation contribute to the compensatory growth after defoliation in the African grass. Defoliation and water stress interact strongly, reducing the effects of defoliation under drought in both native and alien grasses probably due to reduced growth and less transpiring surface in defoliated plants (Simoes and Baruch 1991). The higher defoliation tolerance of the introduced species is possibly related, to their long coevolution with large herbivores in its original African habitat, which did not exist in neotropical savannas for 10 000 years preceding European settlement (Simoes and Baruch 1991).

Although herbivory by native consumers is low in neotropical savannas, the increasing grazing pressure by domestic animals probably plays a role in the dynamics of savanna communities dominated by native and alien grasses and would tend to favor the latter in virtue of their higher defoliation tolerance. However, predation by ants, mainly those of the genera *Atta* and *Acromyrmex*, is one of the main causes of seedling and adult mortality in the alien grasses *M. minutiflora*, *Andropogon gayanus*, and *Brachiaria humidicola* (Klink 1994; Lapointe et al. 1993). At the ecosystem level, higher herbivory tolerance of alien grasses can increase the carrying capacity of the savanna and the population dynamics of the wild and domestic herbivores (Fig. 5.1a). Also, higher herbivory tolerance might facilitate further invasion by alien grasses through the gradual disappearance of less grazing-tolerant native competitors (Fig. 5.1m).

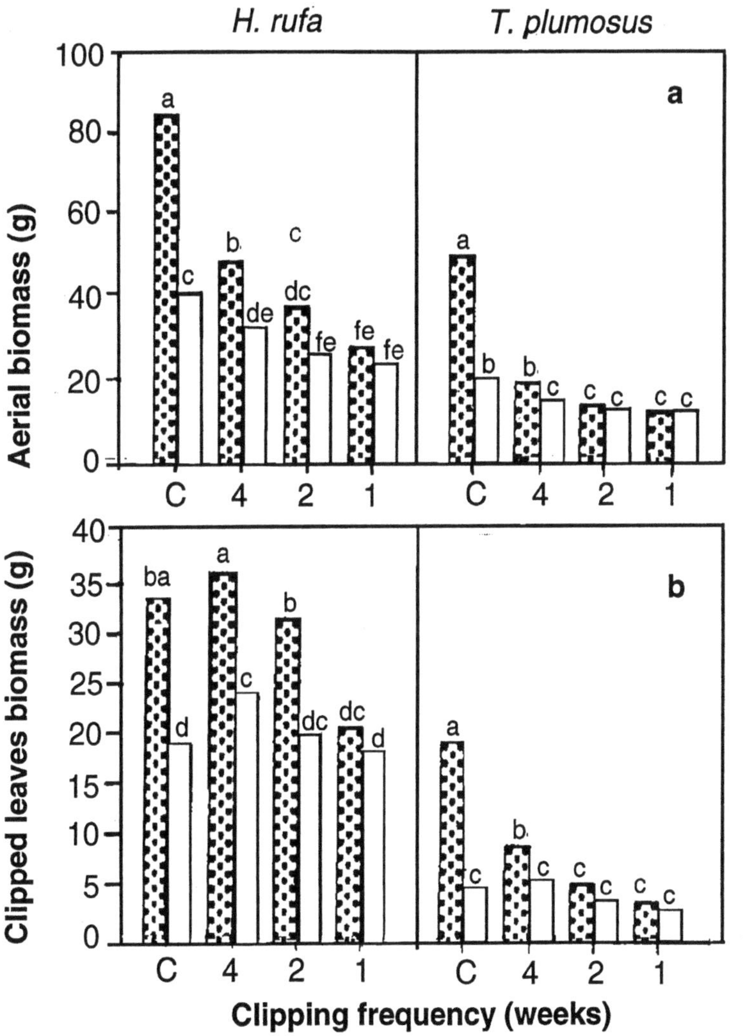

Fig. 5.3. Aerial biomass *a* and clipped leaf biomass *b* under watered (*checkered bars*) and water-stressed (*open bars*) conditions in plants under three clipping frequencies and unclipped controls (*C*). For each species, *bars marked with the same letter* are not significantly different at $P < 0.05$. (Simoes and Baruch 1991)

5.6 Microclimate and Allelopathy

The closed canopy of the communities dominated by alien grasses reduces the intensity and alters the spectral composition of irradiance (decreased red/far-red ratio) that reaches the soil surface (Fig. 5.1g). These microclimatic changes at the soil surface may affect the rate of nutrient cycling through decomposition and inhibit the germination of woody plants altering grass-tree dynamics and probably delaying succession (Fig. 5.1i). However, information about the mechanisms involved is lacking . In addition, some African grasses, such as *M. minutiflora* and *H. rufa*, have allelopathic effects which act either directly, retarding the growth of potential woody competitors (Marinero 1964) or indirectly through the secretion of antibiotics that suppress the growth of nitrifying bacteria (Fig. 5.1k) (Boughey et al. 1964).

5.7 Conclusions

Specific and structural biodiversity has been drastically reduced in neotropical savannas invaded by African grasses. Through the displacement of indigenous species, natural grasslands have been replaced by closed, species-poor, homogeneous stands. There is a reasonable amount of information on the ecophysiology of native and alien grasses and the invasion process. Also the population dynamics of several native and alien grasses has been studied in some detail in Venezuela and Brazil (Silva and Ataroff 1985; Silva and Castro 1989; Silva et al. 1991; Klink 1994, 1995).
There is also considerable information on productivity and nutrient cycling in neotropical savannas (reviewed recently by Sarmiento 1984; Medina and Silva 1990; Medina 1993). However, besides the obvious enhancement of primary production in savannas invaded by African grasses, there is limited information on the effects that biodiversity reductions and increased biomass production have on savanna biogeochemistry. Introduced grasses have the potential to modify all three factors that control the functioning of neotropical savannas: water and nutrient availability, and fire (Frost et al. 1986; Medina and Silva 1990). Therefore, their impact on ecosystem processes such as succession, nutrient cycling, hydrology, and responses to disturbances such as fire and grazing should be significant. Nevertheless, caution is needed in the assessment of the impact of species invasion on ecosystem function, because it is necessary to separate the effects on the competitive processes between populations from the effects that diversity reduction may have on the operation of the ecosystem (Vitousek 1990).

The current tendencies in savanna use aim at the replacement of native communities by cultivated or altered grasslands in order to improve cattle production (Thomas et al. 1990). The process would decrease savanna biodiversity to the same as or a larger extent than does the spontaneous invasion of alien grasses discussed above. Comparative studies on the biogeochemistry of native savannas and those dominated by alien grasses should be undertaken in order to provide the tools for the sustainable management of natural and substituted savannas.

References

Aristeguieta L (1966) Flórula de la Estación Biológica de los Llanos. Bol Soc Ven Cienc Nat 110:228-307

Baker HG (1978) Invasion and replacement in Californian and neotropical grasslands. In: Wilson J R (ed) Plant relations in pastures. CSIRO, Melbourne, pp 368-384

Baruch Z (1986) Comparative ecophysiology of native and introduced grasses in a Neotropical savanna. In: Joss PJ, Lynch PW, Williams OB (eds) Rangelands: a resource under siege. Proc Second Int Rangelands Congress. Adelaide, Australia. Aust Acad Sci, Canberra, pp 449-450

Baruch Z (1989) Ecofisiología de algunos pastos en sabanas tropicales. Monogr Syst Bot Missouri Bot Garden 27:24-36, San Louis, Missouri

Baruch Z, Fernández D (1993) Water relations of native and introduced C_4 grasses in a neotropical savanna. Oecologia 96: 179-185

Baruch Z, Ludlow MM, Davis R (1985) Photosynthetic responses of native and introduced C_4 grasses from Venezuelan savannas. Oecologia 67:288-293

Baruch Z, Hernández AB, Montilla MG (1989) Dinámica del crecimiento, fenología y repartición de biomasa en gramíneas nativas e introducidas de una sabana Neotropical. Ecotropicos 2:1-13

Bazzaz FA (1986) Life history of colonizing plants: some demographic, genetic and physiological features. In: Mooney HA, Drake JA (eds) Ecology of biological invasions of North America and Hawaii. Ecological studies 58. Springer, Berlin Heidelberg New York, pp 96-110

Berendse F, Elberse WT (1990) Competition and nutrient availability in heathland and grassland ecosystems. In: Grace J B, Tilman B (eds) Perspectives in plant competition. Academic Press, New York, pp 93-116

Bilbao B, Medina E (1990) Nitrogen-use efficiency for growth in a cultivated african grass and a native South American pasture grass. J Biogeogr 17:421-425

Boughey AS, Mumo PE, Meiklejohn J, Strang RM, Swift MJ (1964) Antibiotic reactions between African savanna species. Nature 203:1302-1303

Christie EK, Moorby J (1985) Physiological response of semi-arid grasses. I.- The influence of phosphorus supply on growth and phosphorus absorption. Aust J Agric Res 26: 423-436

Cruces J (1977) Productividad primaria, fenología y valor nutritivo de la gramínea *Hyparrhenia rufa* (Nees.) Stapf. en dos localidades del Edo. Guárico. Tésis de Licenciatura, Univ Cent Venez, Caracas, 112 pp

D'Antonio CM, Vitousek PM (1992) Biological invasions by exotic grasses, the grass/fire cycle and global change. Annu Rev Ecol Syst 23:63-87

Frost P, Medina E, Menaut JC, Solbrig O, Swift M, Walker BH (1986) Responses of savannas to stress and disturbance. Int Union Biol Sci Spec Issue 10, Paris, pp 1-82

Hodgkinson KC, Ludlow MM, Mott JJ, Baruch Z (1989) Comparative responses of the savanna grasses *Cenchrus ciliaris* and *Themeda triandra* to defoliation. Oecologia 79:45-52

Klink CA (1994) Effects of clipping on size and tillering of native and African grasses of the Brazilian savannas (the cerrado). Oikos 70:365-376

Klink CA (1995) Germination and seedling establishment of native and invading African grasses in the Brazilian Cerrado. J Trop Ecol (in press)

Lapointe SL, Serrano MS, Villegas A (1993) Colonization of two tropical forage grasses by *Acromyrmex landolti* (Hymenoptera: Formicidae) in Eastern Colombia. Fl Entomol 76:359-365

Marinero RM (1964) Influencia de *Melinis minutiflora* en el crecimiento de *Cordia alliodora*. Turrialba 14:41-43

Medina E (1993) Mineral nutrition: tropical savannas. Progr Bot 54: 237-253

Medina E, Silva JF (1990) Savannas of northern South America: a steady state regulated by water-fire interactions on a background of low nutrient availability. J Biogeogr 17: 403-413
Orians GH (1986) Site characteristics favoring invasions. In: Mooney HA, Drake JA (eds) Ecology of biological invasions of North America and Hawaii. Ecological studies 58. Springer, Berlin Heidelberg New York, pp 133-148
Parsons JJ (1972) Spread of African grasses to the American tropics. J Range Manage 25:12-17
Pieters A (1993) Efecto de la profundidad y fertilidad del suelo sobre el crecimiento, repartición de biomasa y nutrientes de *Hyparrhenia rufa* (Nees) Stapf. Tésis de Licenciatura, Univ Cent Venez, Caracas
Rodriguez JG (1987) Compartamentalización de la fitomasa y del nitrógeno en *Trachypogon plumosus* y *Axonopus canescens*, especies dominantes del pastizal de las sabanas de Trachypogon. Tésis de Licenciatura, Universidad Simón Bolívar, Caracas
San José JJ, Fariñas MR (1991) Temporal changes in the structure of a *Trachypogon* savanna protected for 25 years. Acta Oecol Int J Ecol 12:237-247
San José JJ, Medina E (1976) Organic matter production in the *Trachypogon* savanna at Calabozo, Venezuela. Trop Ecol 17:113-124
Sarmiento G (1984) The ecology of neotropical savannas. Harvard Univ Press Cambridge
Sarmiento G (1992) Adaptive strategies of perennial grasses in South American savannas. J Veg Sci 3: 325-336
Sarmiento G, Monasterio M (1969) Studies on the savanna vegetation of the Venezuelan Llanos. I. The use of association analysis. J Ecol 57:579-598
Silva JF, Ataroff M (1985) Phenology, seed crop and germination of coexisting grass species from a tropical savanna in western Venezuela. Oecol Plant 6: 41-51
Silva JF, Castro F (1989) Fire, growth and survivorship in a neotropical savanna grass *Andropogon semiberbis* in Venezuela. J Trop Ecol 5:387-400
Silva JF, Raventos J, Caswell H (1991) Fire and fire exclusion effects on the growth and survival of two savanna grasses. Acta Oecol 11:783-800
Simoes M, Baruch Z (1991) Responses to simulated herbivory and water stress in two tropical C_4 grasses. Oecologia 88:173-180
Thomas D, Vera RR, Lascano C, Fisher MJ (1990) Use and improvement of pastures in neotropical savannas. In: Sarmiento G (ed) Las sabanas americanas: aspectos de su biogeografía, ecología y utilización. Fondo Editorial Acta Cient Venez, Caracas, pp 141-162
Vareschi V (1970) Las sabanas del Valle de Caracas. Acta Bot Venez 4:427-522
Velásquez J (1965) Estudio fitosociológico acerca de los pastizales de las sabanas de Calabozo, Edo. Guárico. Bol Soc Ven Cienc Nat 109:59-101
Vitousek PM (1990) Biological invasions and ecosystem processes: towards an integration of population biology and ecosystem studies. Oikos 57:7-13

6 Relationships Between Biotic Diversity and Primary Productivity in Savanna Grasslands

6 Relationships Between Biotic Diversity and Primary Productivity in Savanna Grasslands

Luis Bulla

6.1 Introduction

The relationship between plant diversity and ecosystem primary productivity is a strongly debated issue in the ecological literature. Although there is no compelling evidence indicating that higher plant biodiversity should result in higher primary production, it has been frequently assumed that the cooccurrence of different biological forms may increase the efficiency of interception of incident radiation. As a consequence, a number of "humped" curves relating plant diversity to ecosystem production have been documented (Lugo 1988; Tilman 1986; Medina and Huber 1992).

Frequently, the discussion on the relationships between diversity and productivity has been obscured by the heterogeneity and, at times, low reliability of the methods used to measure both parameters. Before I present the other results of the research which constitute the purpose of this chapter, I shall first discuss the methodology applied to assess diversity of primary producers and primary productivity in a given site.

6.2 Diversity Indices

There are a number of diversity indices in the literature, each one with advantages and disadvantages (Magurran 1988)

Ecological Studies, Vol. 121
Solbrig, Medina and Silva (eds) Biodiversity and Savanna Ecosystem Processes

The Shannon-Weaver index

$$H' = \sum p_i ln p_i$$

and its associated measure of equitability (Pielou 1976)

$$J = \frac{H'}{H'_{max}} = \frac{H'}{\log s}$$

are quite common in the ecological literature, but are not convenient ways of diversity assessment (Hurlbert 1971; DeBenedictis 1973; Hill 1973; Peet 1974; Alatalo and Alatalo 1977; Alatalo 1981; Molinari 1989). In part the problems are: (1) the meaning and units of measurement of the H′ index are obscure and change depending on the base of logarithms selected, a fact frequently overlooked by authors using the index without indicating the logarithm base used in the calculations; (2) the index is not additive; (3) it does not have a linear behavior; (4) it is much more sensitive to the equitability of the community than to the number of species; (5) the equitability index associated with the Shannon-Weaver index depends on sample size (DeBenedictis 1973) and overestimates grossly the equitability of very uneven samples.

Indices based on Simpson's dominance measure estimate diversity as (1-C) or (1/C), where

$$C = \sum p_i^2$$

This index emphasizes dominance in a given community, ignoring almost completely the rare species, which are always an important component of diversity in savanna ecosystems. Therefore, it is not satisfactory as measure of biodiversity in savannas.

The indices H' and 1/C are strongly related. Hill (1973) demonstrated that the number of species, the inverse of the Simpson's index, and the exponential of the Shannon-Weaver measure are all different powers of a single mathematical expression, named N_0, N_1, and N_2. The advantages of these numbers are: (1) all are expressed in the same units (number of species); (2) they are additive; and (3) they differ in the weight they give to different species. N_0 (the number of species) gives the same importance to all of them, N_1 (antilog H') emphasizes the importance of the species of intermediate abundance, and N_2 (1/C) gives more importance to the very abundant species.

However, all the diversity indices have a common problem derived from the dual nature of the diversity concept. Diversity of a community encompasses its number of species or "richness", and its "equitability" or "evenness". The diversity indices produce a single number which is a weighted combination of both factors; but this can lead to serious mistakes. A community with very few species but a high equitability may have the same diversity index as a community with many species but very low equitability. It is assumed then, that both magnitudes can somehow compensate one another, which obviously does not make any biological sense. Therefore, for best results, species richness and community equitability should be maintained apart when analyzing the relationship between diversity and primary productivity.

6.3 Species Richness in Savanna Vegetation

When a complete census of a community is available, it provides a perfect measure of its richness. However, that is seldom the case. The usual procedure is to sample a certain number of quadrats. If two workers sample a different number of quadrats their results will not be comparable. Sander's rarefaction method as modified by Hurlbert (1971) could be applied to predict the number of species to be found in a sample of n individuals, rarefactioned from a larger sample size N. Although the technique looks promising, further investigation is required in order to solve two main problems: (1) usually in savannas the abundance of a given species is characterized by its percent cover or frequency and not by the number of individuals; (2) the method assumes that the presence of a given species does not modify the probability of occurrence of any other, a condition that does not apply to savanna grasslands.

6.4 Measurement of Equitability

The equitability index of Pielou (1976) is quite common in the ecological literature, but it frequently overestimates equitability, and is strictly nonlinear (Molinari 1989).

The indices derived from Hill's numbers E_{21} (Hill 1973) and F_{21} (Alatalo 1981), or the modification of F_{21} introduced by Molinari (1989) are being increasingly used but have several problems. The differences in behavior of the various equitability indices described here can be shown with a community constituted by only two species (Molinari 1989); (Fig. 6.1). F_{21}, E_{21}, and J are strictly nonlinear over this gradient of only two species, but Molinari (1989) G_{21} seems to perform better.

However, all equitability indices described so far and especially F_{21} and G_{21} do not have a clear ecological interpretation (Bulla 1994) and are strongly biased in the importance they give to the different species in the

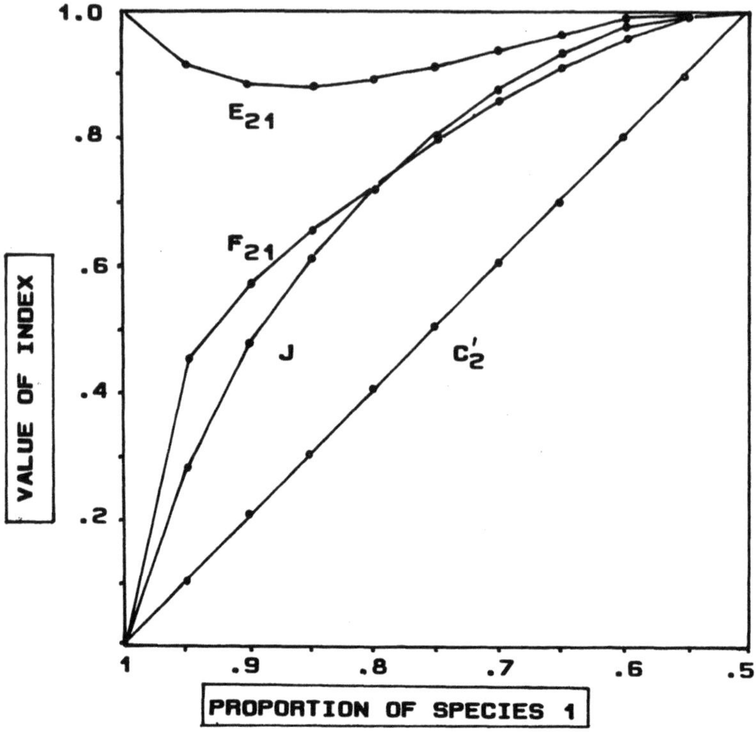

Fig. 6.1. The values of several equitability indices along a gradient of increasing equitability

sample. These indices almost ignore species with very low frequency values, therefore, these species contribute little to the actual index value; but in all savanna grasslands occurrence of rare species is quite common. In a large set of sites in Venezuela, more than 40% of the species have abundances well below 1%. Therefore, a different equitability measurement has to be used to account for this large proportion of rare species in the community, an index which gives equal weight to all species. This index has been derived by Bulla (1994). Its basic idea is shown by means of the following example (Fig. 6.2).

In Fig. 6.2 the species have been ordered according to their frequencies, and are compared with the theoretical distribution of a community of the same richness but perfect equitability, where all species have the same frequency. This proportion pi will be

$$p_i = \frac{1}{S}$$

where S= number of species.

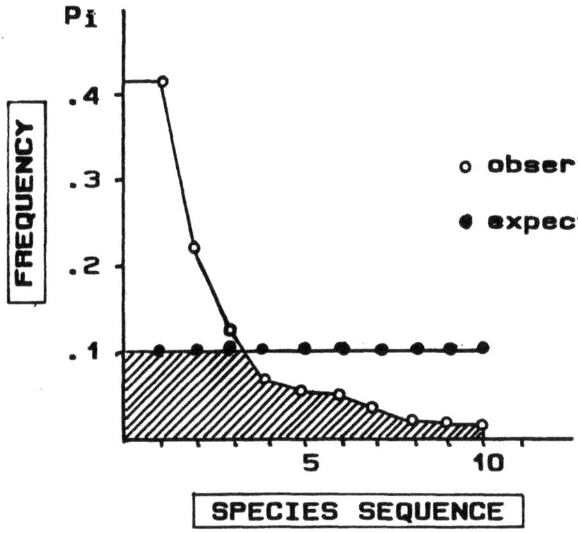

Fig. 6.2. The relative abundances of ten species found in a "ripio" savanna, and the expected theoretical distribution for a community with the same number of species but perfect evenness. The *shaded area* represents the overlap between the two distributions

The degree of overlap between the observed (pj) and the theoretical (pi) distribution is a perfect measure of the equitability of the sample. The best measurement of overlap between observed and expected proportions is the Czekanowski's proportional similarity index (Feisinger et al. 1981) also known in the ecological literature as the Schoener's index of niche overlap:

$$O = 1 - 0.5\sum |p_i - p_j| = \sum^{\mathrm{MIN}(p_i, p_j)},$$

where min (pi, pj) indicates the minimum of the observed and the expected proportions and the sum is over all species in the sample. In order to have an index of equitability based on O and which varies between 0 and 1 the following formula may be used:

$$E = \frac{0 - (1/S)}{1 - (1/S)}$$

This index will be used later to assess species evenness of grasslands ecosystems in Venezuela (Bulla 1994) A simple diversity measure may be calculated as D=ExS, where S is the number of species in the sample, and D the diversity measure.

6.5 Measurement of Primary Productivity of Grasslands

Primary productivity of grasslands can be conveniently and accurately measured using a simple five compartment model (Bulla et al. 1980, 1981, 1990; Fig. 6.3).

All the above-ground biomass of the system is allocated to one of the following compartments: green biomass (GREEN), standing dead biomass (DEAD), and litter (LITTER). Accumulation and transfer of plant organic matter between the compartments occurs through production (PRODU), mortality (MORT), decomposition of standing dead matter (DECOM), fall of dry matter (FALL), and litter decomposition (DECOL). The behavior of the system through time is described by a set of difference equations:

GREEN (T) = GREEN (T-1) - MORT (T-1,T) + PRODU (T-1,T)
DEAD (T) = DEAD (T-1) + MORT (T-1,T) - DECOM(T-1,T) - FALL (T-1,T)
LITTER (T) = LITTER (T-1) + FALL (T-1,T) - DECOL (T-1,T).

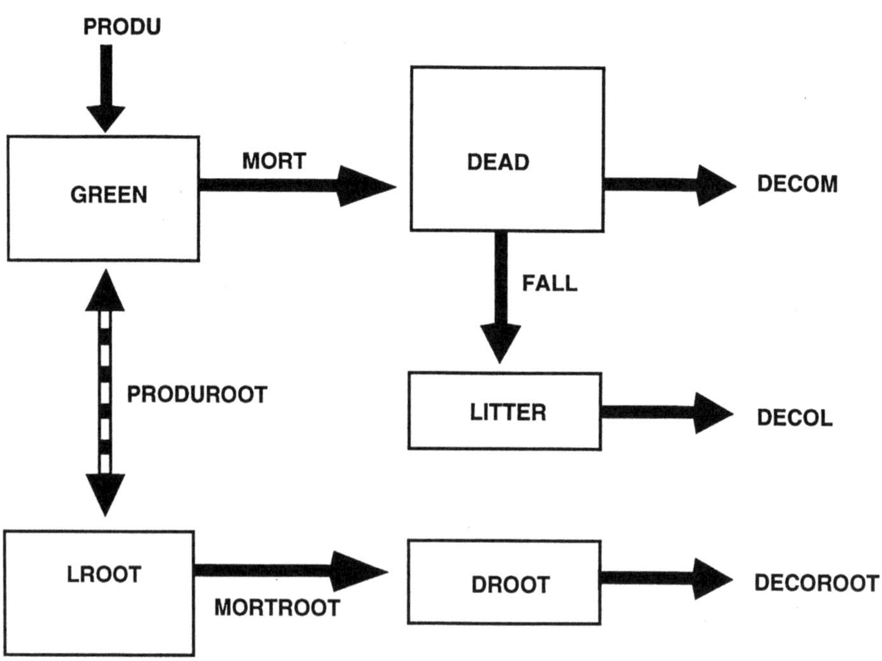

Fig. 6.3. The compartment model proposed by Bulla et al. (1981) for the measurement of primary production

In a similar way, a set of two equations may be formulated to describe the changes in biomass taking place in the roots compartment. All below-ground biomass is assigned either to live roots (LROOT) or dead roots (DROOT). Transfer between compartments is described by the following difference equations:

LROOT (T) = LROOT (T-1) + PRODUROOT (T-1,T) - MORTROOT(T-1,T)
DROOT (T) = DROOT (T-1) + MORTROOT (T-1,T) - DECOROOT (T-1,T)

Measuring the size of the different compartments, and the rate of litter (and dead root) decomposition, the system can be solved for all the transfer rates, including production of organic matter. Some characteristics of this model should be emphasized:

1. It determines the minimum information required for an accurate estimation of primary productivity using harvest methods. Any simplification of the estimation will produce large errors (Bulla et al. 1980; Kinyamario and Imbamba 1992).
2. It is not necessary to measure all the quantities specified in the model. For the above-ground part at every sampling date we estimate only the amounts of GREEN, DEAD and LITTER material. We estimate simultaneously DECOM and DECOL using litter bags. The remaining three parameters, MORT, FALL, and PRODU are calculated solving the system of equations.
3. The system is solved for each time interval. It allows the calculation of the production, mortality, and fall that took place during each interval between subsequent sampling dates. Annual productivity is calculated summing the monthly productivities during a full year.
4. The model provides a detailed picture of the biomass fluxes in the system.

The model is conceptually identical with the one used by Long and Jones (1992) in several grassland ecosystems. A detailed account of the advantages and limitations for the application of this model to grasslands is given in Bulla et al. (1981) and Long and Jones (1992).

When the information required to run the model described above is not complete, an alternative to estimate total annual above-ground productivity in grasslands is to measure the amount of green + dead biomass during the period when peak green biomass is reached (approximately during the middle of the rainy season). This amount is better correlated with net primary above-ground productivity than green biomass alone (Fig. 6.4). Consequently, all subsequent analyses on the relationships between productivity and plant diversity will be based on estimations of above-ground productivity using peak total standing crop (green + dead), which is available for a great number of savannas.

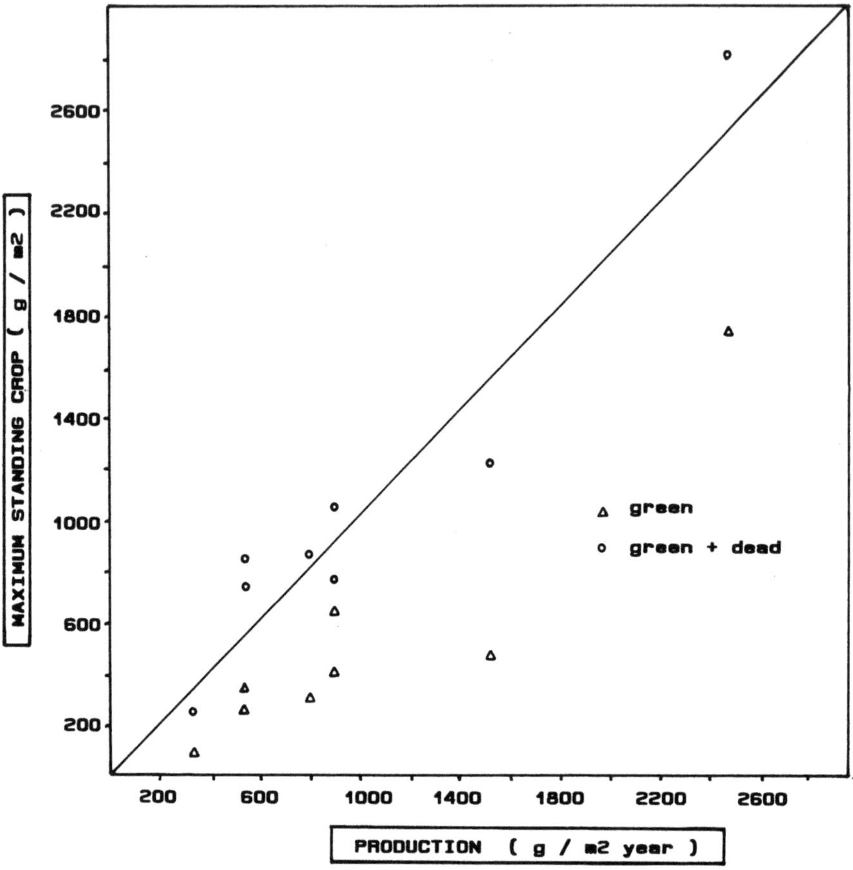

Fig. 6.4. The relationship between productivity and peak standing crop of green or total (green + dead) material. The *line across the figure* represents the ideal situation of a perfect estimation. All *points below the line* represent underestimations, *over it* overestimation. Data from Bulla et al. (1980, 1990) (Venezuela); Garcia-Moya and Montanez (1992) (Mexico); Kinyamario and Imbamba (1992) (Nairobi); Kamnalrut and Evenson (1992) (Thailand)

6.6 Relationships Between Productivity and Diversity

The relationship between productivity (peak standing crop) and diversity (estimated by the total number of species found in 25 samples of 1 m^2) was investigated in 36 grassland sites in Venezuela (Fig. 6.5). The data set includes sites from the Gran Sabana (Bolivar State), the central plains (Llanos in Guarico State), northern Amazonas State, montane secondary grasslands near Caracas, coastal grasslands (Falcon State), and a few others. The set covers a large range of biomass, from less than 150 to more

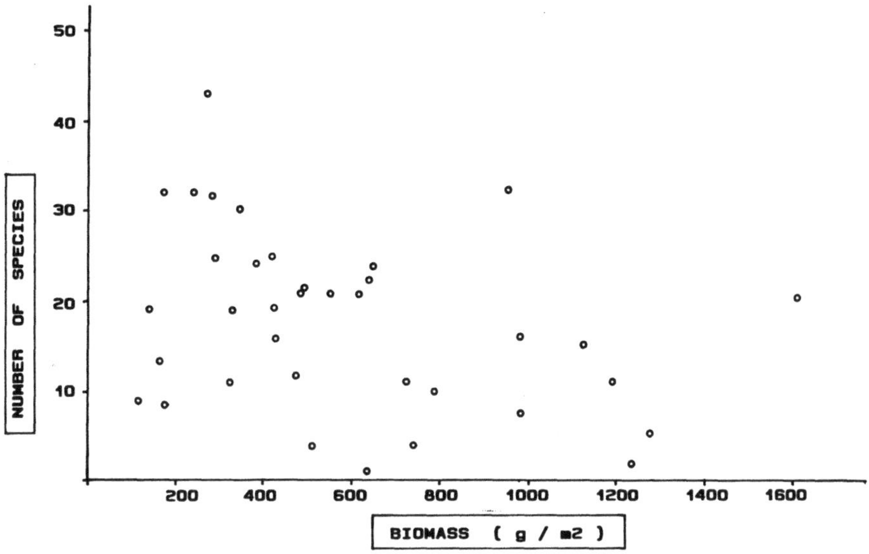

Fig. 6.5. The relationship between peak standing crop and number of species for 36 Venezuelan savannas

than 1600 g m^{-2}, and species numbers ranging from 10 to 46. Both variables appear to be independent of each other, since no clear relationship is observed. There are some grassland sites with biomass values between 100-300 g m^{-2} where the number of species varies from 9 to 46, and sites with biomass larger than 800 g m^{-2} with species numbers ranging from 2-33. This result simply highlights the fact that there are many factors affecting the diversity of each area, thereby obscuring its relation with productivity.

To minimize the effects of these uncontrolled variables and to simplify the analysis, we will restrict the following presentation to the comparison of grasslands located in the same area, under the same general conditions. Two groups were selected, one from the Gran Sabana area (15 sites) and another from de Central Plains (6 sites).

A brief description of the methodology used to collect the data is required, because most of these data sets have not been published before. Diversity was measured by means of a census of species found in 25 quadrats of 1 m^2. In each quadrat all species were identified and their relative cover was visually estimated. To estimate green to dead biomass ratios, 10 additional quadrats of 0.25 m^2 area were harvested. Sampling included 21 grasslands sites, resulting in a grand total of 525 diversity samples and 210 biomass samples. Based on these data I will analyze the behavior and relationships of the following parameters:

1. Total number of species found for each grassland in the 25 samples, and the mean number of species per sample.
2. The pattern of relative frequencies of the different species in each grassland.
3. The total diversity of each grassland measured by the new D index described before, as well as H′, N, and N_2.
4. The equitability of each grassland measured by the new E index discussed previously, as well as F_{21} and J.
5. The proportion of the total diversity of each grassland attributable to within samples and between samples diversity (only for the Central Plains sites).
6. Biomass and the green/dead ratios.

The analysis of each grassland series will clarify the interactions between productivity (as inferred from peak biomass) and the different components of diversity, and the mechanisms involved.

Table 6.1. Biomass, richness, evenness and diversity indices for four savannas of the Yuruani series

	Savanna Number			
	1	2	3	4
No. of spp.	18	30	30	23
Biomass g/m^2	135	268	340	659
Evenness indices				
J	0,906	0,893	0,857	0,888
F21	0,832	0,834	0,803	0,786
G21	0,521	0,524	0,477	0,452
E	0,633	0,608	0,486	0,478
Diversity indices				
H'	2,62	3,04	2,92	2,78
N1	13,7	20,9	18,5	16,2
N2	11,6	17,6	15,0	12,9
D	11,4	18,2	14,6	13,3

6.7 Gran Sabana Grasslands

The Gran Sabana is a large area located in the southeastern corner of Venezuela. Detailed descriptions of floristic and geographic characteristics of the area can be found in Huber (1986). All sites sampled were characterized by a rolling hills topography, and it was found that the location of grassland sites along the topographic gradient greatly influenced their productivity and floristic composition. The 15 grasslands sites examined in this section are conveniently separated in three groups each one corresponding to a topographical sequence:

1. San Francisco de Yuruani at 950 m above sea level, four sites from the top to the bottom of a small hill.
2. Luepa I, 1100 m asl, four sites selected as above from the top to the bottom of small hills.
3. Luepa II, 1100 m asl, seven sites located along a 2-km long transect.

6.8 Topographical Gradients

In Yuruaní, the plot on the hill top is the poorest in species with only 18 species, while those in intermediate positions had 30 species, and that on the base of the hill had 23 species (Table 6.1 and Fig. 6.6). The grasslands in intermediate positions had the highest amount of rare species, while the grassland on the top, that grows on a gravel substrate (ripio) had a higher equitability. As expected, biomass (g m^{-2}) increased steadily from 135 g/m^2 at the top of the hill, to 268 g/m^2 and 340 g/m^2 in intermediate positions, and to 659 g/m^2 at the base of the hill. Equitability (E) decreased slowly from the hill-top to the hill-base (from 0.63 to 0.48) , while diversity was higher in the intermediate quadrats (18.2 and 14.6) compared to both hill-top (11.4) and hill-base (13.3) plots. The same pattern was observed in the Luepa topographical gradient.

The relationship between biomass and the number of species or diversity indices is not linear (Fig. 7.6A, B), but shows higher values at intermediate levels of biomass production. If productivity of a grassland site is considered to be positively correlated with resource availability (or inversely correlated with environmental stress), our results could be taken as a confirmation of the hypothesis that systems at the extremes of the resource availability axis tend to be less diverse than the intermediate ones (Tilman 1986; Medina and Huber 1992). The hill-top grasslands seems to be stress-dominated, with low diversity and biomass, but high equitability. The hill-base grassland, with better water and nutrient availability, and developing

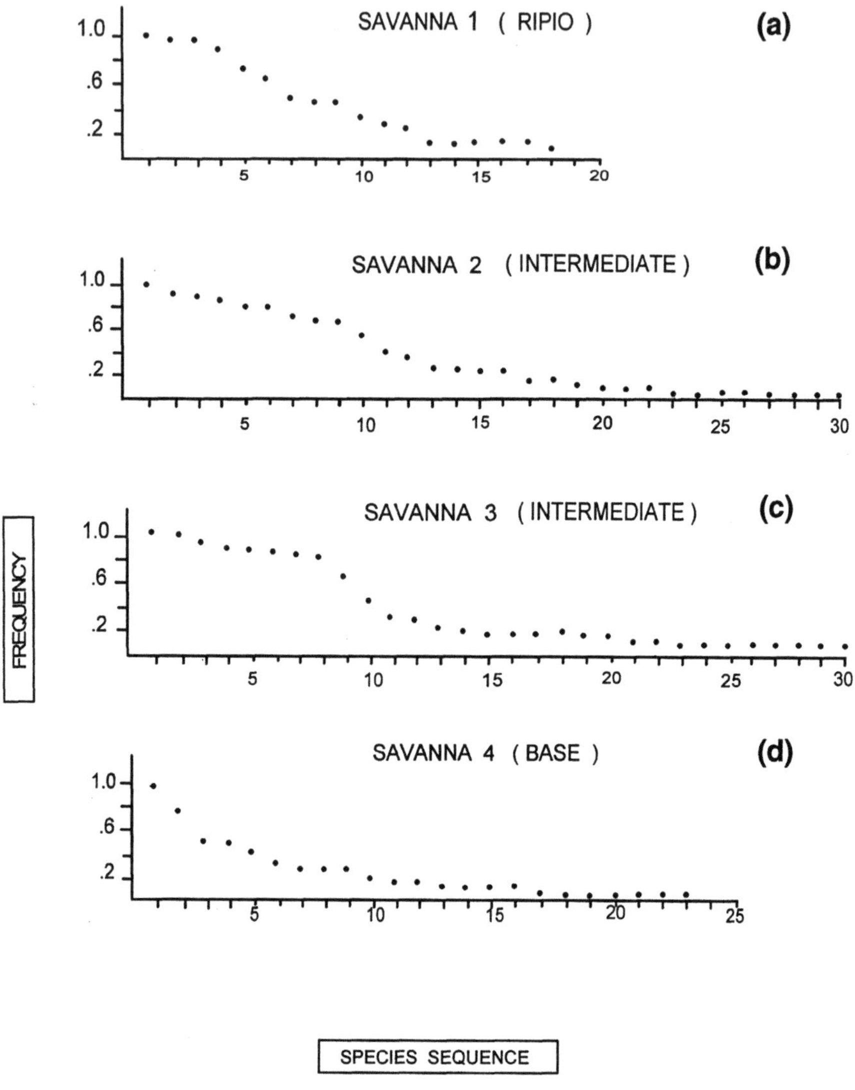

Fig. 6.6a-d. The relative frequencies of all the species found in the four savannas of the Yuruani series. **a** Savanna 1, the "ripio" savanna in the top of the hill. **b** and **c** Savannas 2 and 3 in an intermediate position in the slope. **d** Savanna 4 at the base of the hill. For further explanation see text

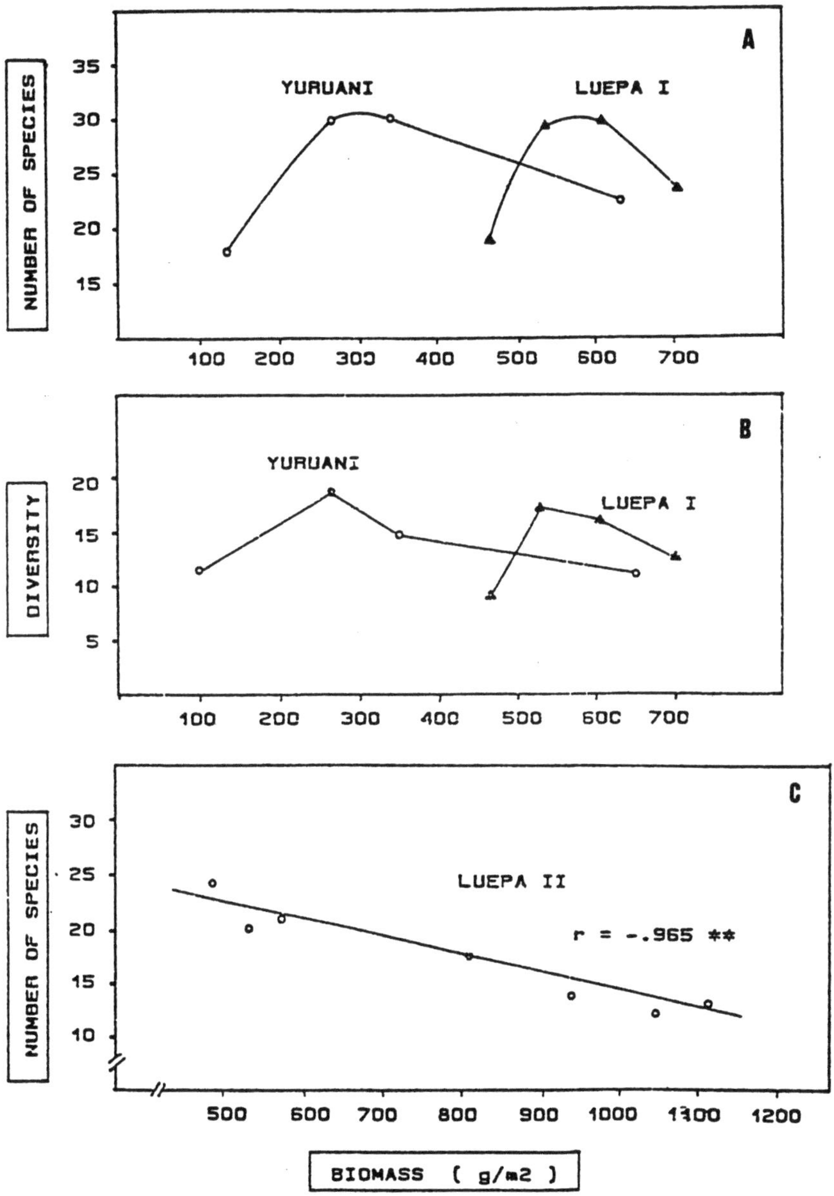

Fig. 6.7. The relationship between biomass (g/m^2) and A number of species or B diversity, measured by Bulla's (1994) D index for the Yurani and Luepa I topographical sequences. C The relationship between biomass and number of species for the seven savannas of the Luepa II gradient. The correlation coefficient is highly significant ($P < 0.01$)

the largest biomass in the gradient, seems to be competition-dominated. Its total cover reached 85% compared to 28 and 45% in the hill-top and intermediate grasslands, but its equitability is much lower. The hill-base grassland is an impoverished version of the intermediate ones, with a few species showing high abundance and a large number with very low frequencies. This is the expected distribution of relative abundances in a competition-dominated community.

The grasslands in Luepa I gave a similar pattern, although the species (or diversity) vs. biomass curve is displaced to the right, because their biomass is much larger (Fig. 6.7A, B).

The larger transect measured in Luepa II, covering a biomass range from 450 to 1100 g m^{-2} showed a clear reduction in species number as biomass increased (Fig. 6.7C), and seems to represent only the right section of the more general Luepa I and Yuruani; curves.

We conclude that:

1. Within a certain vegetation type, there is not a linear relationship between biomass and diversity, but maximum diversity is reached at intermediate values of productivity. Reductions in diversity appear to be related to stress at the lower end of productivity, and to competition at the higher end of productivity.
2. Communities under higher stress and lower productivity have higher equitability than competition-dominated high productivity ones. The reason is that in stressful environments tolerant species are selected for, and productivity potential is low. Therefore, successful species have enough space, competition is low, and each species may reach a roughly equivalent population size, resulting in higher equitability. In competition-dominated environments, the most productive species occupy most of the space, reducing population size or eliminating the less competitive ones, thus leading to lower equitability.

6.9 The Calabozo Series

The six grasslands of the Calabozo series are located in the savannas of the Estación Biológica de los Llanos (Sociedad Venezolana de Ciencias Naturales, Calabozo, Guárico State). These savannas have been well studied, and several reports on their floristic composition, phenology, productivity, root dynamics, soils, etc. have been published (among many others: Blydenstein 1962, 1963; Ramia 1967; San José and Medina 1975; Sarmiento 1983). The savanna inside the Estación Biológica has been protected from fire for the last 25 years, thus producing important changes in many of its characteristics (Fariñas and San José 1987).

Table 6.2. Biomass, number of species, evenness and diversity indices for three pairs of savannas of the Calabozo series

	Savanna 1 (poor)		Savanna 2 (intermediate)		Savanna 3 (rich)	
	Burned	Protected	Burned	Protected	Burned	Protected
No. of spp.	16	10	32	21	25	21
Biomass g/m^2	280	480	424	720	425	1653
Evenness indices						
J	0.880	0.758	0.834	0.821	0.884	0.777
F21	0.774	0.649	0.712	0.663	0.754	0.438
E	0.620	0.400	0.430	0.400	0.570	0.320
Diversity indices						
H'	2.44	1.74	2.84	2.50	2.84	2.36
N1	11.50	5.70	18.00	12.20	17.20	10.60
N2	9.10	4.10	13.10	8.40	13.20	5.20
D	10.00	4.00	13.70	8.40	14.20	6.70
Diversity within and between samples (%)						
Within	31.3	24.0	24.4	20.0	33.1	10.7
Between	68.7	76.0	75.6	80.0	66.9	89.3

In this section we compare sites of paired grasslands that have been either protected against fire or burned regularly during the last 25 years. Grasslands pairs are located side by side, separated by a fence and a 10-m-wide bare soil strip used as a fire break. Otherwise, soils and topographic positions were similar. The three grassland pairs selected differed in their productivity. The first pair (low production) (Table 6.2) was located on a partially disintegrated lithoplinthyte soil, a ferruginous hard-pan near the soil surface that reduces soil volume and makes plant establishment difficult. The second (intermediate production) and third (high production) (Table 6.2) pairs were located in deeper soils, but the soil of the third pair had higher nutrient availability (data not shown). All grasslands were sampled at the peak of the rainy season.

In each pair the number of species in the protected (nonburned) grassland was lower than in the burned one (Fig. 6.8). This reduction is due to a decline in the abundances of the intermediate and rare species, leading to the disappearance of some of them. The reduction in species richness amounted to 37, 34 and 19% in the low, intermediate, and high production grasslands, respectively. The number of species in burned plots was lower in the low (16) and the high production grassland (26) than in the intermediate production grassland (32).

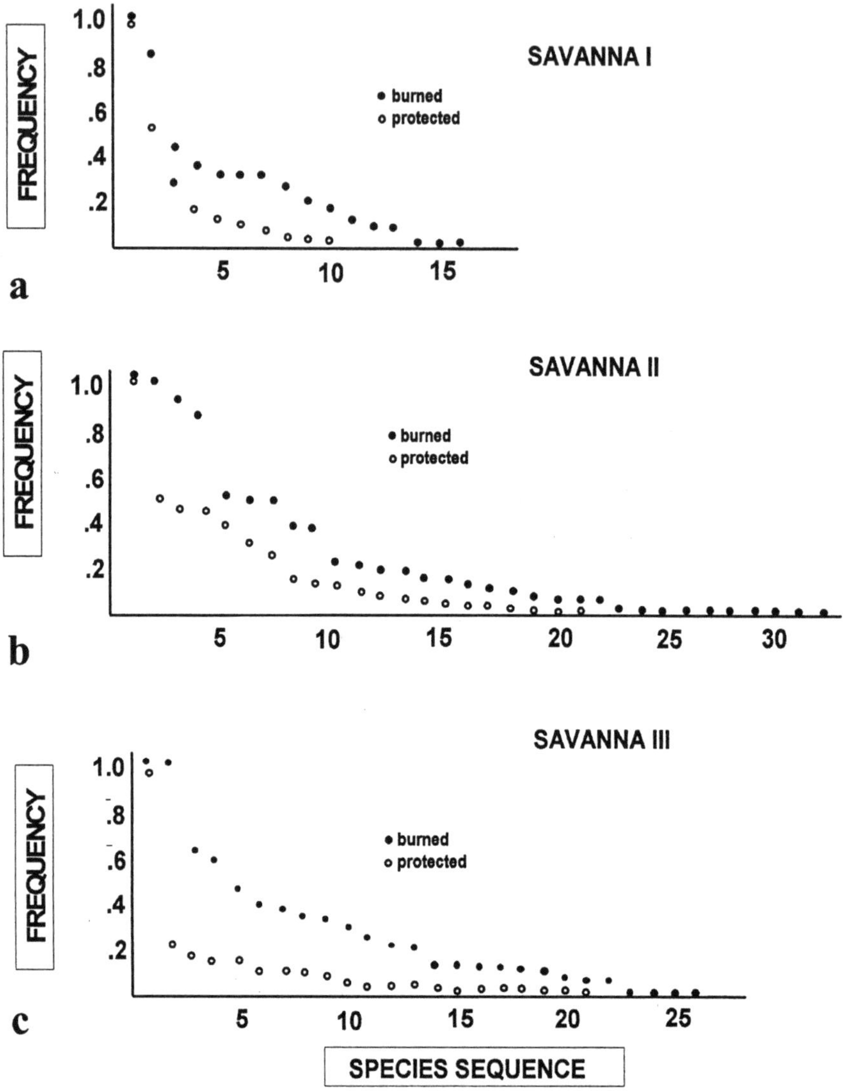

Fig. 6.8 The relative frequencies of all species recorded for each of the three pairs of savannas of the Calabozo series. *Each graph* represents a protected and a burned area. **a** Savanna I the poor "ripio" savanna. **b** The intermediate production savanna II. **c** The high production savanna III

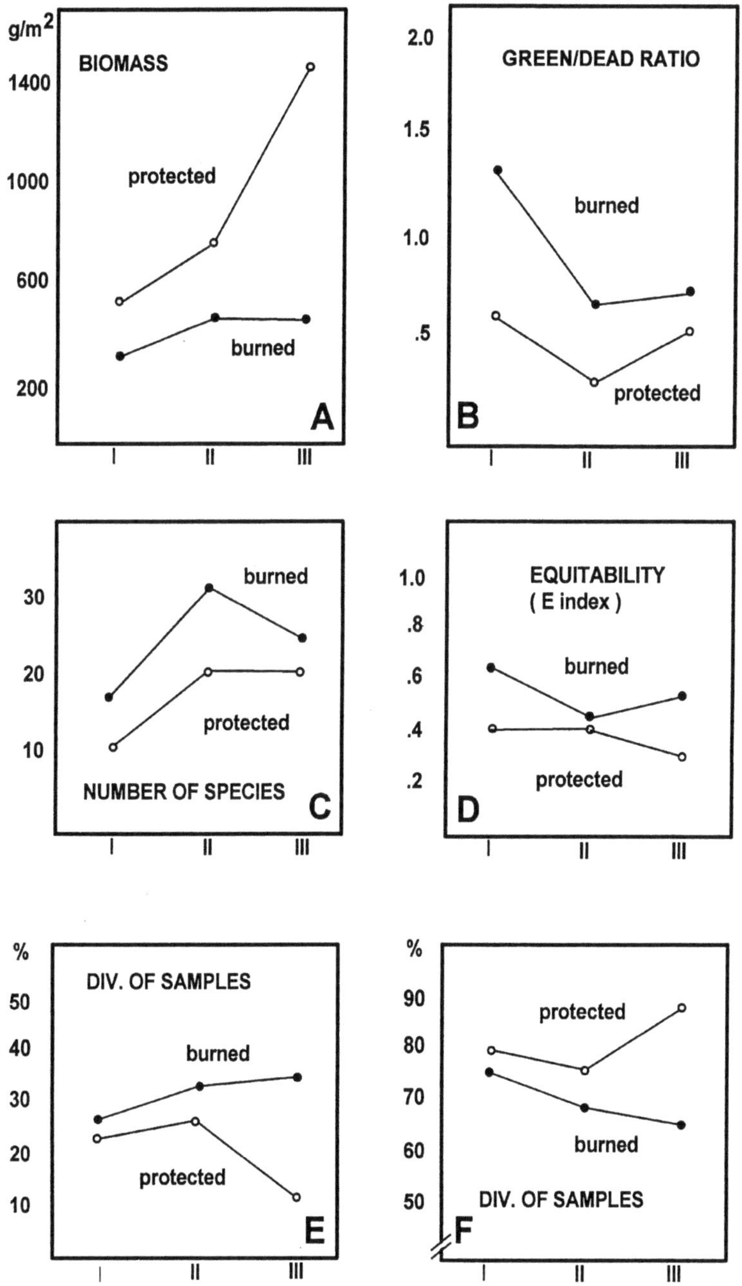

Fig. 6.9 A-F The changes in several characteristics of th savannas of the Calabozo series produced by the protection against fire; *I* is the poor savanna and *III* is the rich one

The protection against fire resulted in a substantial increase in the standing crop of these grasslands in direct proportion to their productivity (Fig. 6.9A). Not surprisingly, but as expected, the green/dead ratio was much larger in the regularly burned grasslands (Fig. 6.9B).

Interestingly, fire protection results in a sharp drop in both equitability and number of species, indicating that the elimination of fire disturbance favors the development of dominant species and leads to species disappearance (Figs. 6.9C and D).

An interesting aspect of the fire effects that can be observed in this data set is that mean diversity of individual samples, expressed as a fraction of the total diversity of each grassland, is much higher in burned sites, while the diversity among samples, possibly a measure of spatial heterogeneity, is much higher in protected sites (Fig. 6.8E, F). This means that protection against fire creates a community with fewer species, but spatially more heterogeneous. Such a pattern results when some species are eliminated by competition from some patches but remain in others.

The previous results may be explained on the basis of the increase in biomass accumulation that takes place when fire is eliminated. These results contradict previous reports indicating that long-term protection against fire increases diversity of the grass layer (Fariñas and San José 1987). However, our observations correspond to one-time comparison of plots maintained under different fire regimes, while Fariñas and San José (1987) followed changes in biodiversity through time.

Concluding remarks on the relationships between diversity and productivity:

1. Diversity is the result of the interaction of multiple causes, productivity being only one of them, and not necessarily the most important. Therefore the relationships between these two variables may appear weak, or even nonexistent when many sites are compared. The only way out is to carefully select comparable communities so as to eliminate other factors of disturbance.
2. The accurate estimation of primary production requires the measurement of several variables preferably at monthly intervals, and the use of the methods of Long and Jones (1992) or Bulla et al. (1990). This methodology has been applied only to a relatively small number of savannas, and cannot be used at this time for the purpose of this chapter. Given this restriction, the best point estimator of the above-ground production of grasslands ecosystems seems to be the maximum total standing crop (green + dead) measured at the peak of the rainy season or when the green biomass reaches its peak.
3. When studied along gradients of increasing productivity (peak standing crop) the relation between production and diversity follows a curvilinear pattern, with minimum diversity associated with both high or low biomass and maximum at intermediate values. The position of this

"intermediate" point of inflexion is not fixed and depends on the characteristics of the zone. For example, in Yuruaní it is around 300 g m^{-2}, while in Luepa it is in the order of 500-600 g m^{-2}.

4. The equitability component of diversity has an inverse relation to biomass. Poor communities tend to be more equitative than highly productive ones.
5. Our results confirm the hypotheses of Tilman (1986), Lugo (1988) and Medina and Huber (1992) that both low productivity, stress-driven communities and high productivity, competition dominated ones will have lower diversities than intermediate ones. In the first case, the size of the populations will be resource-limited, and competition may be low. In the second case, resources are in larger supply and those species capable of trapping them faster, or more efficiently, will be more productive and tend to occupy all the space available, therefore competition may be high. In both cases, the number of species will be comparatively low, but the two communities are structured by different forces. In addition, it is not clear how multiple stresses interact in determining occurrence of species assemblages of variable tolerance.
6. The elimination of fire produces an accumulation of biomass that reduces diversity of the system. The effect is proportionally stronger in the stress-dominated savanna. There is also a sharp decline in the equitability of the community
7. The new equitability index E seems to be a more sensitive tool than other indices most commonly used in the literature.

References

Alatalo R (1981) Problems in the measurement of evenness in ecology. Oikos 37:199-204
Alatalo R, Alatalo A (1977) Components of diversity: multivariate analysis with interaction. Ecology 58:900-906
Blydenstein J (1962) La vegetación de la Estación Biológica de los Llanos. Bol Soc Venez Cienc Nat 101:97-136
Blydenstein J (1963) Cambios en la vegetación despues de la protección contra el fuego. Bol Soc Venez Cienc Nat 103:223-238
Bulla L (1994) A new index of evenness and its associated diversity measure. Oikos 70:167-171
Bulla L, Miranda R, Pacheco J (1980) Producción, flujo de materia orgánica y diversidad de una sabana de banco del Módulo Experimental de Mantecal. (Venezuela). Acta Cient Venez 31: 31-38
Bulla L, Pacheco J, Miranda R (1981) A simple model for the measurement of primary productions in grasslands. Bol Soc Venez Cienc Nat 136:281-304
Bulla L, Pacheco J, Morales G (1990) Sesonally flooded neotropical savanna closed by dykes. In: Bremeyer A (ed) Managed grasslands. Elsevier, Amsterdam, pp 177-211
DeBenedictis A (1973) On the correlation between certain diversity indices. Am Nat 107:295-302
Fariñas M, San José JJ (1987) Efectos de la supresión del fuego y del pastoreo sobre las composición de una sabana de Trachypogon de los llanos occidentales. In: San José JJ, Montes R (eds) La capacidad bioproductiva de sabanas. Cent Int Ecol Trop Unesco-IVIC, Caracas, pp 513-545
Feisinger P, Spears E, Poole W (1981) A simple measure of niche breadth. Ecology 62:27-32
Garcia Moya E, Montanez P (1992) Saline grassland near Mexico City. In: Long S, Jones B, Roberts J (eds) Primary productivity of grass ecosystems, Chapman & Hall, London New York, pp. 70-99
Hill MO (1973) Diversity and evenness: a unifying notation and its consequences. Ecology 54: 427-432
Huber O (1986) La Gran Sabana y su ambiente natural. In: Todtmann O (ed) La Gran Sabana. Caracas pp 159-179
Hurlbert SH (1971) The non-concept of species diversity: a critique and alternative parameters. Ecology 52:577-585
Kamnalrut A, Evenson J (1992) Monsoon grassland in Thailand. In: Long S, Jones M, Roberts M (eds) Primary productivity of grass ecosystems. Chapman & Hall, London New York, pp 100-126
Kinyamario HI, Imbamba SK (1992) Savanna at Nairobi National Park, Nairobi. In: Long SP, Jones MB, Roberts MJ (eds) Primary productivity of grass ecosystems. Chapman & Hall. London, pp 25-65
Long SP, Jones MB (1992) Introduction, aims, goals and general methods. In: Long SP, Jones MB, Roberts MJ (eds) Primary productivity of grass ecosystems. Chapman & Hall, London, pp 1-24
Lugo A (1988) Diversity of tropical species: questions that elude answers. Biol Int Spec Issue 19:37, IUBS, Paris
Magurran AE (1988) Ecological diversity and its measurement. Princeton Univ Press, Princeton, 179 pp
Medina E, Huber O (1992) The role of biodiversity in the functioning of savanna ecosystems. In: Solbrig OT, van Emden HM, van Dordt PGWJ (eds) Biodiversity and global change, Monogr 8. Int Union Biol Sci, Paris, pp 139-158
Molinari J (1989) La diversidad ecológica, un enfoque unificado conceptual y metodológico para su cuantificación. Tesis Doctoral, Univ Cent Venenz, Caracas, 164 pp
Peet RK (1974) The measurement of species diversity. Annu Rev Ecol Syst 5:285-307

Pielou EE (1976) Mathematical ecology. Wiley, New York, pp 385
Ramia M (1967) Tipos de sabanas de los llanos de Venezuela. Bol Soc Venez Cienc Nat 27:264-288
San José JJ, Medina E (1975) Effect of fire on organic matter production and water balance in a tropical savanna. In: Golley FB, Medina E (eds) Tropical ecological systems. Springer, Berlin Heidelberg New York, pp 251-264
Sarmiento G (1983) Patterns of specific and phenological diversity in the grass community of the Venezuelan tropical savannas. J Biogeogr 10:373-391
Simpson EH (1949) Measurement of diversity. Nature 163: 688
Tilman D (1986) Resources, competition and the dynamics of plant communities. In: Crowley MJ (ed) Plant ecology. Blackwell, Oxford, pp 51-75

7 Biodiversity and Fire in the Savanna Landscape

7 Biodiversity and Fire in the Savanna Landscape

R. W. Braithwaite

7.1 Introduction

Biodiversity or biological diversity is, as Solbrig (1992) points out, not an entity or resource but a property, a characteristic of nature. Species, populations, certain kinds of tissues are resources, but not their diversity as such. Diversity is a defining characteristic of life. In order to deal with it scientifically, we need standard ways of measuring it. Biodiversity has been operationally defined as a hierarchical product of genetic, species, and community (or ecosystem) diversity (Walker and Nix 1993). Each level is ecologically significant and each can be accorded a value in its own right. Similarly, Groombridge (1992) reports the widespread practice of defining biodiversity in terms of genes, species, and ecosystems. The present chapter considers the three levels, though I know of no specific research on the genic level in tropical savannas. The only approximation available seems to be the degree of genetic distinctiveness represented by the level of endemicity.

The Responses of Savannas to Stress and Disturbance group of ecologists has long identified four major determinants of tropical savannas as plant-available moisture (PAM), available nutrients (AN), fire and herbivory (Frost et al. 1986; Chap. 2). The first two are regarded as the primary determinants and the latter two are secondary modifiers of savannas (Solbrig 1991). There is much to be learned about biodiversity in relation to the PAM-AN plane and herbivory, a project we are getting going along the North Australia Tropical Transect (NATT) (Braithwaite et al. 1992). The present chapter attempts to assemble what is known about the influence of fire on the biodiversity of tropical savannas.

Ecological Studies, Vol. 121
Solbrig, Medina and Silva (eds) Biodiversity and Savanna Ecosystem Processes

I draw very heavily on Australian examples. In particular, experiments at Munmarlary, Katherine, and Kapalga in the Northern Territory. The first two were established in the 1970s with plots of about a hectare and primarily addressed vegetation issues. The Kapalga study was only established in 1990 and has large catchment-sized (15-20 km^2) treatment areas. It is being used to address a wide range of questions. All three experiments are in areas with a very low level of grazing by native (Press 1988a) or introduced mammals as a consequence of the nutrient-poor nature of the herbage (Mott et al. 1981; Cook and Andrew 1991).

7.2 Plants

7.2.1 Woody and Herbaceous

Much has been written about savanna plants and fire. However, there is not so much written about diversity in relation to fire. Typically, areas of savanna protected from fire show a steady increase in species, particularly rain forest species (Gillon 1983; Medina and Huber 1992; Fig. 7.1). The West African example shows a much slower rate of increase than the Venezuelan one. In Australia, an even slower succession or no succession is seen. While almost all of northern Australia would be core rather than derived savanna, Braithwaite (1991) has also argued that resulting from the historical and environmental circumstances of the Australian continent, the Australian savanna biota is not so closely affiliated with its rain forest biota as it seems to be on other continents.

At Munmarlary in the Northern Territory, an assessment of the impact of four experimental treatments was done after 5 years (Hoare et al. 1980). The species richness of under-story and mid-story plants had changed little, with some individual plots increasing slightly and others decreasing slightly. Mean number of species in open forest were 14 and 13 for annual early season fire, 17 and 16 for annual late season fire, 15 and 13 for biennial fire, and 14 and 13 for no fire for 1973 and 1978 respectively. No pattern is discernible. After 13 years, there was still no significant difference between treatments in woodland but in open forest significant differences ($p<0.05$) were obtained with the highest mean value of 35.4 species being obtained for late annual, the most intense fire regime. The other means were 30.4, 31.7, and 29.9 for early annual, early biennial, and unburnt treatments respectively. Concurrently, very substantial structural changes took place in the unburnt sites, with massive growth of the mid-layer (Fig. 7.2; Bowman et al. 1988b).

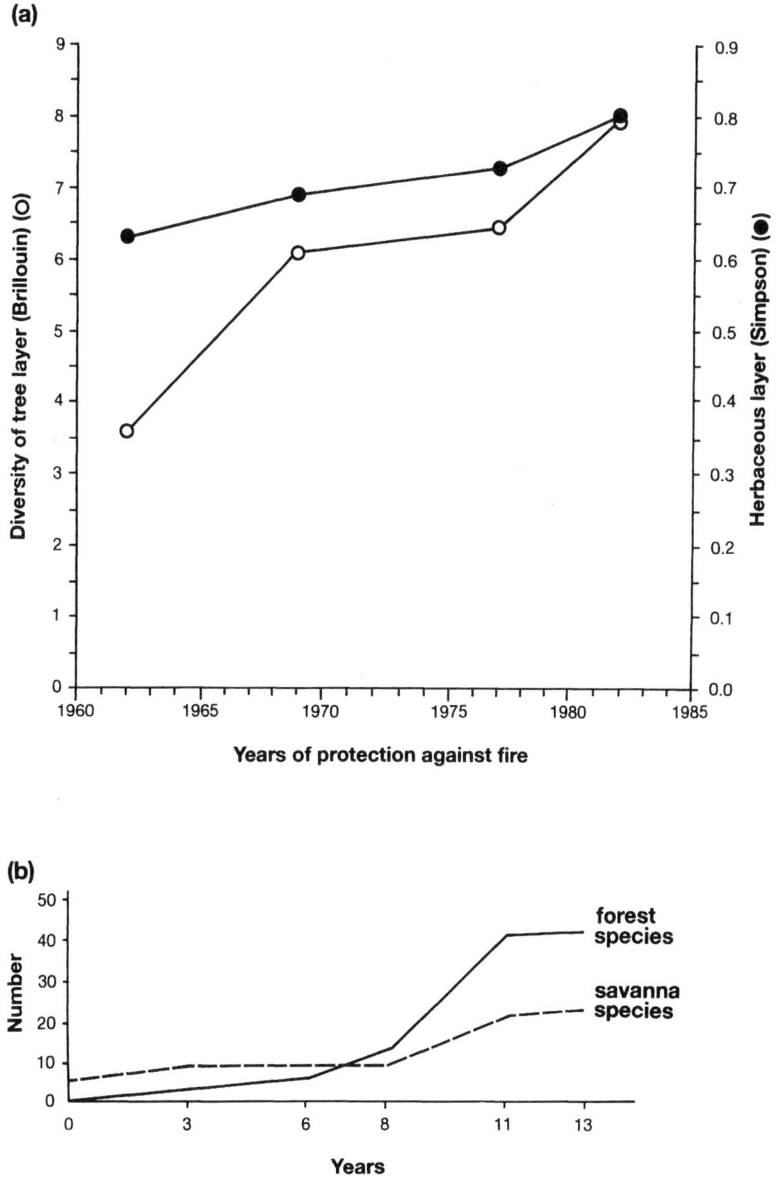

Fig. 7.1.a Variation in diversity of diversity of tree and herbaceaous layer in a 3-ha savanna plot protected from fire since 1961 at the Biological Station at Calabóso, Venezuela. (After Medina and Huber 1992). **b** Change in numbers of woody species per hectare following complete protection from burning in the derived savanna of Lamto, Ivory Coast. (After Menaut in Gillon 1983)

At Solar Village near Darwin, Fencham (1990) compared sites unburnt for 10 years with nearby sites with a known history of burning. He obtained mean species richnesses of 32.5, 41.2, 39.3 per 100 m^2 on unburnt, infrequently but lightly burnt, and annually burnt sites, respectively ($p<0.001$). While there was no evidence of rain forest species in the unburnt sites at Munmarlary, there was at Solar Village with a number of rain forest edge species being recorded (Fencham 1990), a process which appears to be continuing, particularly downslope of Fencham's study sites (J. Brock, pers. comm. 1993). The Munmarlary sites are much more remote (ca. 5 km) from rain forest source areas than the Solar Village sites (ca. 2 km).

It should be remembered that the rain forest in the Northern Territory is depauperate in plant species compared with the local savanna, with counts of ca. 40 species per 0.16 ha quadrats in rain forest (monsoon forest) versus counts of ca. 60 for savanna (open forest and woodland) sites (Taylor and Dunlop 1985). So a loss of species would be expected to be associated with the transition to rain forest in the Northern Territory. This issue is taken up again below under habitat diversity.

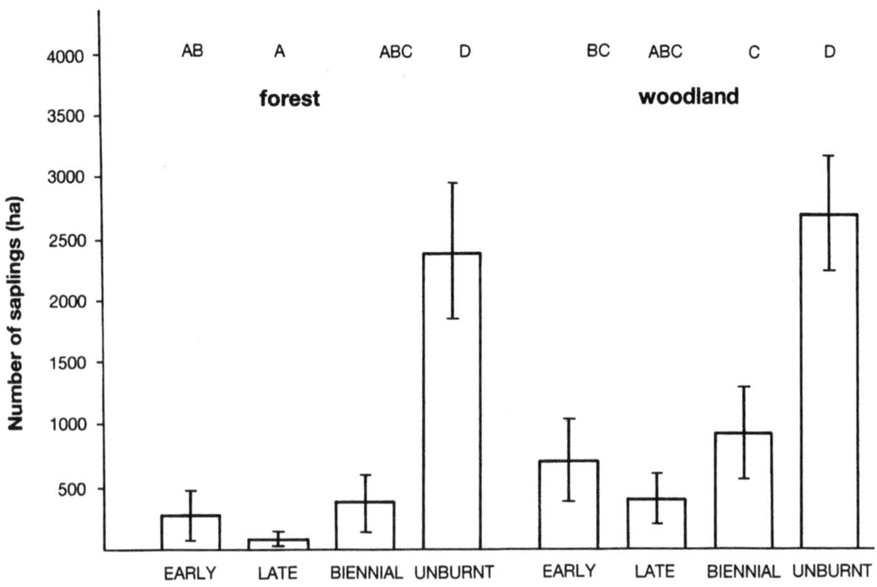

Fig. 7.2. Histogram showing pooled density of tree and shrub saplings greater than 2 m high and less than 10 cm DBH for early annual, late annual, early biennial, and unburnt treatments at Munmarlary, Northern Territory of Australia. *Vertical bars* represent 95% confidence intervals. (After Bowman et al. 1988b).

7.2.2 Epiphytes

After 16 years of the Munmarlary experiment, the epiphytic orchid *Dendrobium affine* was present on all four treatments, but the distribution was extremely uneven. The extent of colonization of suitable trees was 20 times greater on the unburnt treatments but there were no significant differences between the three burning treatments (Cook 1991).

7.3 Habitat Diversity

Habitat diversity is recognized as an important component of biodiversity. As alluded to above, absence of fire can cause a change of state from savanna to rain forest. Conversely, fire can erode the edges of rain forest and other habitats, converting them to savanna (Stott et al. 1990). Fire can change the mix of habitat types by increasing the dominance of one type over others, thereby diminishing biodiversity. In other situations, it may be used to increase biodiversity by increasing habitat diversity. However, I am not advocating that such maximization of biodiversity should be the only consideration in managing land for conservation.

In the Northern Territory, small patches of rain forest (monsoon forest) are associated with localized areas of permanent moisture and locations topographically protected from fire (Bowman et al. 1991; Russell-Smith 1991). It is clear that fire enables eucalypt savanna to push back the boundaries of rain forest patches (Bowman et al. 1988a) and can eliminate rain forest entirely from sites (Bowman and Wightman 1985). Conversely, as has been found elsewhere (Kellman 1984), in the absence of fire, rain forest can invade the savanna (Bowman and Fencham 1991). However, the process is extremely slow in Australia, even in the wet tropics. Nonetheless, it is a threat to the actual persistence of wet eucalypt associations which are found in close proximity to the Queensland rain forest (G.N.Harrington, pers. comm. 1992).

The rocky sandstone escarpment habitats of northern Australia possess many endemic species, both plants and vertebrates (Freeland et al. 1988; Bowman et al. 1988a). The closed forests dominated by *Allosyncarpia ternata* and open forest rich in *Callitris intratropica* are both escarpment habitats which appear to be vulnerable to fire (Bowman et al. 1988a; Bowman 1991; Russell-Smith et al. 1993). The argument is that intense fires from the savanna move up into the escarpment and expose the habitat there to hotter fires than had been usual historically under Aboriginal management. It is likely that there are other escarpment habitats which are being changed into savanna by fire. The heath-land type of vegetation which occurs on the sandy outwash areas at the base of the escarpment might be such a case.

7.4 Insects

7.4.1 Ants

Andersen (1991) has analyzed the guild structure of ant communities on three fire treatments at Munmarlary after 14 years. Ant communities on annually burned plots were characterized by relatively large numbers of individuals of the dominant *Iridomyrmex* group, hot climate specialists, and opportunistic *Rhytidoponera aurata,* and low numbers of individuals of generalized myrmicines and of cryptic species. The reverse was true for unburned plots with generalized myrmicines and cryptic species in high numbers and the other previously mentioned groups much less abundant. The biennial sites showed a pattern of guild structure intermediate between the two patterns above. Many individual species were common under one fire regime but were rarely or never recorded under another. The differences were attributed to structural changes in the habitat caused by fire, and in particular the level of litter accumulation and insulation on the ground. These changes influenced ants directly but also through the dominant *Iridomyrmex* changed the competitive environment. The annually and biennially burned sites had about twice as many species per trap as the unburnt sites.

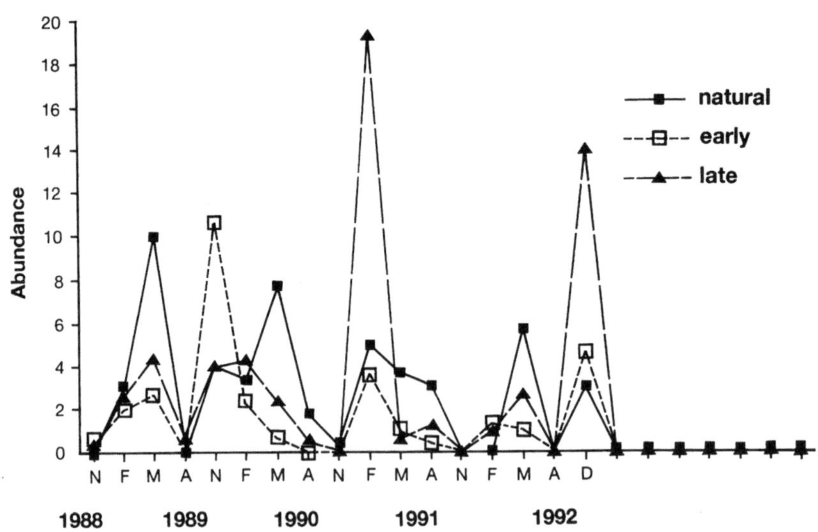

Fig. 7.3. Abundance (mean number of individuals per 150 sweeps) of lepidopteran larvae for three experimental fire regimes at Kapalga in Kakadu National Park, Australia. (A.N.Andersen, unpubl.)

7.4.2 Lepidopterans

Sweep net samples on three regimes of the Kapalga fire experiment showed a changed abundance of caterpillars after the fires commenced in 1990 (Fig. 7.3). Peaks of abundance were obtained in the wet seasons of 1990-1991 and 1992-1993 on the Late sites (A. N. Andersen, unpubl. 1993). This is likely to be the result of nutritious canopy regrowth after the intense late fires. The lack of a late peak in 1991-1992 may be related to the lower mean fire intensity of 8300 kW/m, compared with 12 700 and 9700 kW/m in the other 2 years. These results show that a fire type widely regarded as unequivocally destructive is beneficial to some species. It is also likely that the increase in these herbivorous insects would increase food resources for some insectivorous bird species.

7.5 Vertebrates

Many authors have suggested that fires or changes in fire regime are responsible for the decline in particular species of vertebrates. It has almost become a universal explanation when other explanations are not obvious. However, hard data are few.

7.5.1 Herpetofauna

Using an approach involving the monitoring of the ambient fire regimes and faunal abundance on 54 sites in Kakadu National Park, Braithwaite (1987) found different species to be most abundant under different fire regimes. No one regime, including absence of fire, was best for all or most species. Species richness of species for lizards which breed in the dry season (arid zone phylogenetic affinities) was highest under early dry season patchy fires. The species richness of those which breed in the wet season (wet tropical affinities) were most strongly associated with ground vegetation conditions during the wet season.

Trainor and Woinarski (1994) subsequently examined the same suite of species using the more rigorous methodology of pitfall trapping within the Kapalga fire experiment, and obtained a similar range of responses by different lizard species.

For amphibians in the Kapalga fire experiment, numbers of records increased from 31 to 56 in the Early (annually burnt in early dry season), from 31 to 37 in the Progressive (three successive burns down a progressively drying moisture gradient over the dry season each year), from 39 to 44 in the Late (annually burnt late in each dry season), but decreased from

131 to 74 in the Natural (unburnt since 1987), thereby implying that burning was beneficial to frog populations. The pattern of species richness was one of increase in the three burned treatments and negligible change in the Natural sites from 2 years pre-fire to 2 years post-fire (L.K. Corbett, unpubl. 1993).

7.5.2 Birds

In addition to the birds which hunt around fires and scavenge immediately afterwards, many bird species are attracted to areas that have been recently burnt (Braithwaite and Estbergs 1987; Woinarski 1990; Fig. 7.4). These are mostly granivorous, omnivorous, and carnivorous species which feed on the ground. Influxes of these species occurred because fires increased accessibility to food by clearing the ground of its previously dense ground cover. Studies in the Northern Territory show that the succession of bird species to long-unburnt areas is relatively limited. Although species which fed or nested in the shrubby understory occurred at greater densities in such areas, there was no pronounced change in bird species composition resulting from fire exclusion. Frugivorous species may, however, become more abundant in areas protected from fire, thereby very slowly driving change to monsoon (rain) forest. The cooler early dry season fires appear more beneficial to granivores than the hot late dry season fires (Woinarski 1990).

In the Darwin area, Crawford (1979) has demonstrated greater species richness of ground and shrub layer and canopy-using birds in burnt savanna. For lower strata, this was especially true for dry season migrants. Total species richness for all birds was 9.1 and 11.8 species per census of 1.6 ha plots on unburnt and burnt sites, respectively.

In subtropical Queensland near Maryborough, fires are less frequent and birds responded to different regimes but apparently mainly to structural differences, specifically the proportions of grass and shrubs (Porter and Henderson 1983). Regular observations along transects in plots burnt annually, every 2-5 years, and protected from fire for 29 years showed common use by five, three, and six different resident species respectively. Another 15 common species showed no preference.

7.5.3 Mammals

At Kapalga, 12 species of mammals from 10 g to 2.5 kg in size, were regularly live-trapped. In total, the numbers of individuals changed little post-fire, but on the Natural sites, numbers increased by 250% in the second year after the commencement of the fire experiment. However, mean species richness decreased on Progressive but increased on the other three treatments in the dry season. During the wet season, no change was seen on

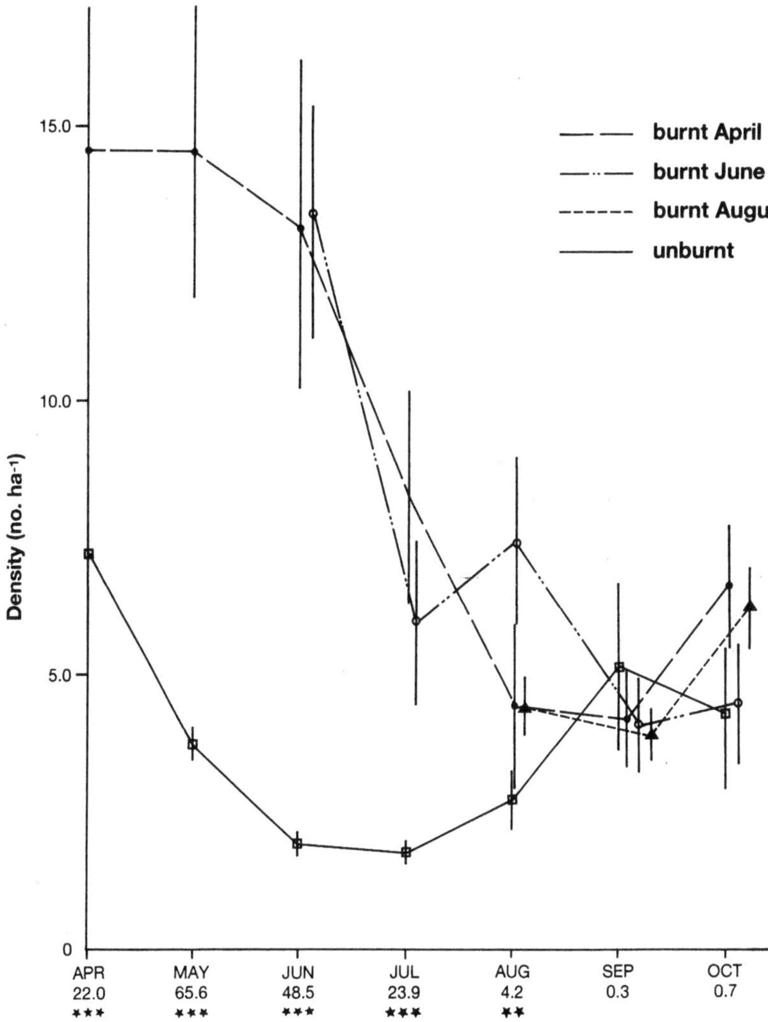

Fig. 7.4. Monthly changes in density of all bird species combined for Katherine study site. *Vertical lines* represent one standard error. *Values under months* are F-ratios, with significance indicated: *** $p<0.001$, ** $p<0.01$, * $p<0.05$. (After Woinarski 1990)

Natural but substantial decreases were seen on the other three treatments. However, no clear and significant pattern was seen with diversity overall (Fig. 7.5). The preferences of individual species often changed between the 2 years of burning in the experiment (Table 7.1). As the productivity, flowering, and fruiting rates of different food plants varies between fire treatments and years (G. D. Cook, R. J. Williams, R. W. Braithwaite, unpubl.), it is likely that the very substantial year-to-year variation in the distribution and

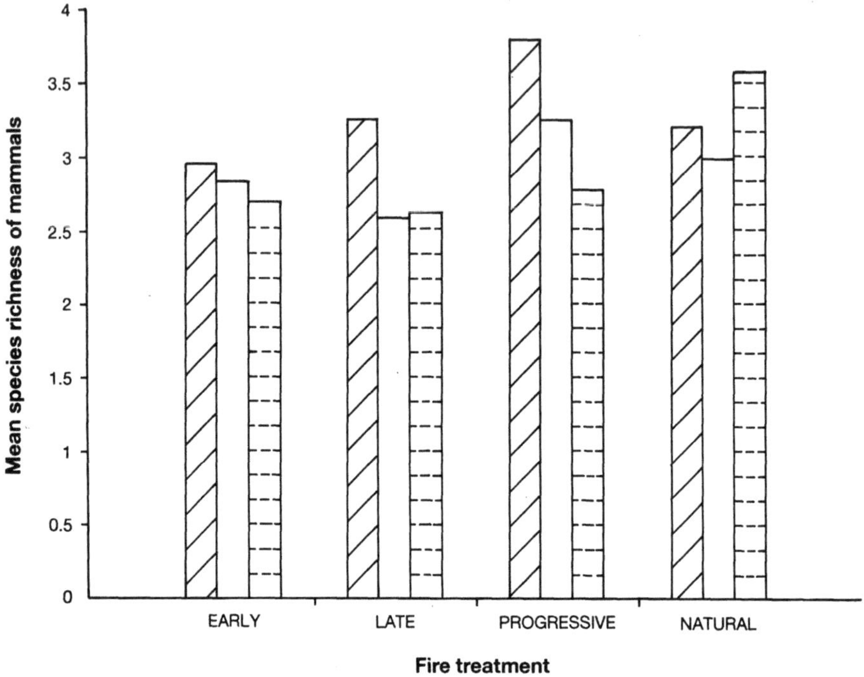

Fig. 7.5. Mean species richness of small mammals for 3 years under four fire treatments at Kapalga, Kakadu National Park, Australia. *Oblique lined bars* represent 1989-1990 pre-fire data, *open bars* 1990-1991 post-fires year, and *horizontal lined bars* 1991-1992 post-fires year. (R.W.Braithwaite, unpubl.)

Table 7.1. Top preference by mammals of the four fire treatments at Kapalga in the 2 years following the commencement of the experiment

Species	Year 1	Year 2
Dasyurus hallucatus	Natural	Natural
Antechinus bellus	Late	Natural
Isoodon macrourus	-	Natural
Trichosurus vulpecula	Early	Natural
Melomys burtoni	Progressive	Natural
Rattus colletti	-	-
Rattus tunneyi	Natural	Early

abundance of rainfall (Taylor and Tulloch 1985) interacts strongly with fire time and intensity to affect the abundance of food plants in this species-rich community (Taylor and Dunlop 1985).

Relationship Between Components of Biodiversity

At a broad community level, Specht (1990) has demonstrated strong correlations between number of mammal species and number of plant species for both tropical northern and mediterranean southern Australia (Fig. 7.6).

Regions of greater habitat diversity (for topographic/geological reasons) are often richer in species of animals (i.e., alpha diversity; Woinarski 1990; Braithwaite 1990). Whether habitat diversity can be manipulated with fire to produce increased animal diversity has never, to my knowledge, been tested.

In addition to the habitat diversity discussed above, there is the within-habitat diversity, or spatial heterogeneity, or patchiness within the savanna. Unlike in Africa (Belsky 1986), herbivory is not a major force on Australian sites and thus is not a major force generating heterogeneity. From Kapalga data, there are significant correlations between mammal species richness and vegetation (floristic) patchiness (Beta diversity) for both ground and canopy vegetation after 2 years without fire before the fire experiment commenced (Fig. 7.7). However, after 2 years of burning, the ground layer relationship had disappeared and the canopy layer relationship was weakened. Obviously there are short-term effects of the fires and these differ

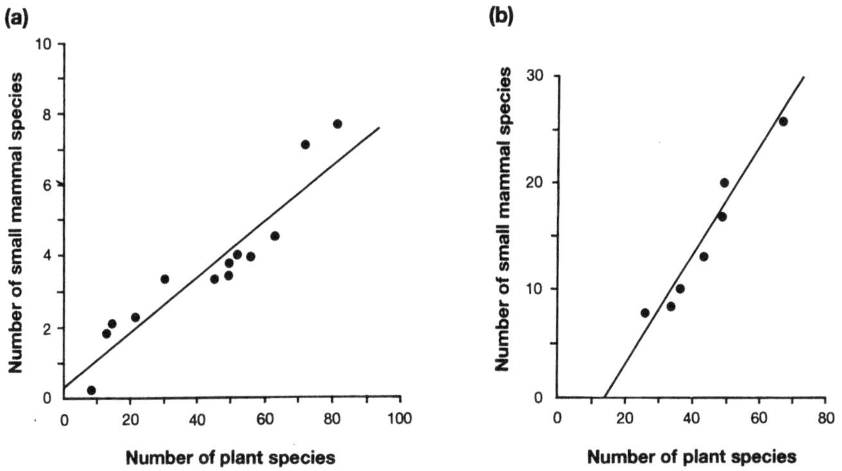

Fig. 7.6a.b. Number of small mammals recorded in plant communities in tropical Australia and in mediterranean southern Australia plotted against number of plant species recorded in the same ecosystems. a Tropics. b Mediterranean. (After Specht 1990)

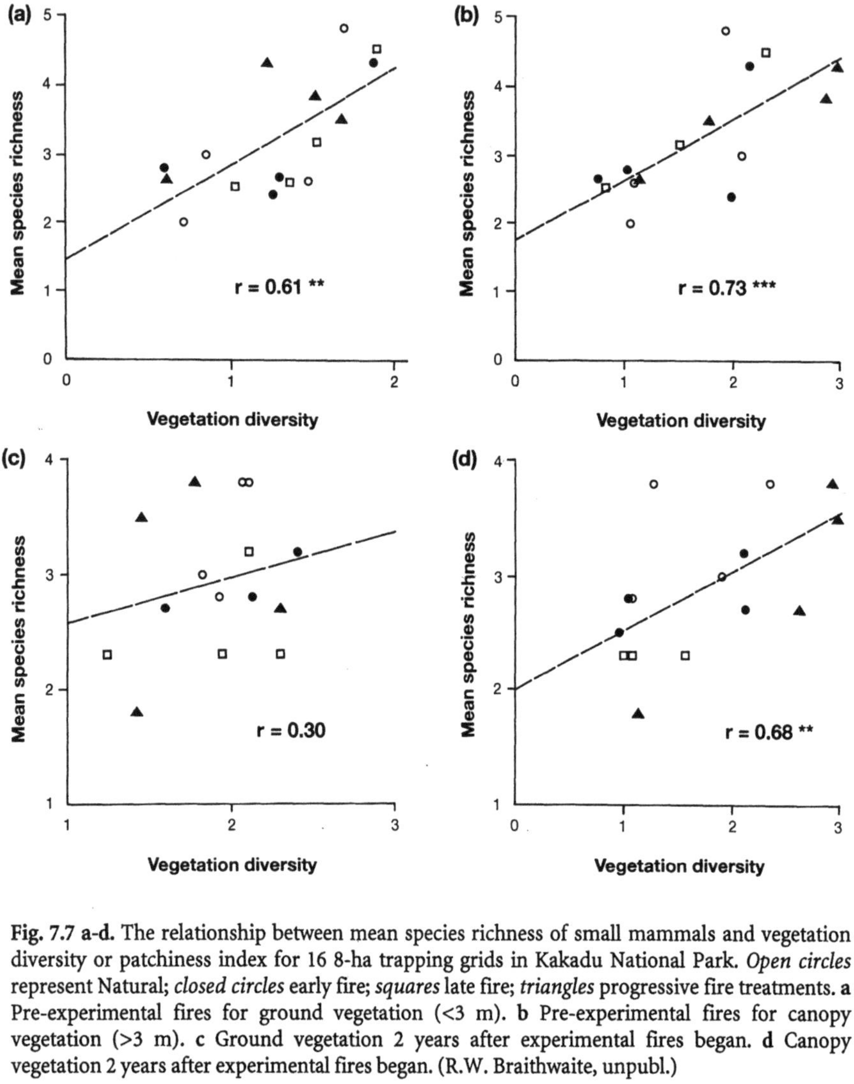

Fig. 7.7 a-d. The relationship between mean species richness of small mammals and vegetation diversity or patchiness index for 16 8-ha trapping grids in Kakadu National Park. *Open circles* represent Natural; *closed circles* early fire; *squares* late fire; *triangles* progressive fire treatments. **a** Pre-experimental fires for ground vegetation (<3 m). **b** Pre-experimental fires for canopy vegetation (>3 m). **c** Ground vegetation 2 years after experimental fires began. **d** Canopy vegetation 2 years after experimental fires began. (R.W. Braithwaite, unpubl.)

between fire treatments (unpubl.). It would now be interesting to cease the fire treatments and see if particular fire treatments had improved the relationships of the immediate pre-fire situation. The impacts of fire are also complex. The impact of fire intensity on the phenology of a single species appears relatively simple, but when the range of life history responses from the savanna species is considered, the collective outcome of the interaction with fire becomes complex. How can the savanna be burned in so many different ways yet basically remain the same ? How does the patchiness of the savanna relate to the species diversity of the biota ?

7.6 A Model of Savanna Homeostasis

Competition between the canopy and ground layers of vegetation has long been a dominant theme for savanna ecologists (Walter 1971). While climatic and edaphic factors are the major determinants, the balance of dominance can move from one stratum to the other, with fire and herbivores a common agent of change (Belsky 1990). In Australia, fire with some facilitation from wood-eating termites is the main cause of tree loss (ca. 70% of total), with windthrow (ca. 10%) and lightning (ca. 10%) and termites (ca. 10%) as minor agents (Braithwaite 1985), and absence of fire the main facilitator of recruitment to the canopy (Hoare et al. 1980). The concept of "clump-interclump" (Hoare et al. 1980) or fine-scale patchiness is one of the defining elements of savanna; trees and grass are a spatial manifestation of this. However, the patchiness within each of the ground and canopy layers is also a typical savanna characteristic which both determines and is determined by fire (Stott 1986).

Conventional wisdom, at least in Australia, has been that the early dry season fires are patchy on the ground and impact negligibly on the canopy, thereby protecting canopy resources. This is seen as the desirable management goal and is often claimed to be *the* Aboriginal burning regime (Jones 1980). These early dry season fires typically burn only part of the ground cover within a patch of savanna (Kapalga mean=73.8%, range= 30-99%, n=9), leaving many individual small plants untouched by fire (Braithwaite and Estbergs 1985). However, as the young plants are still actively growing at this time of year, they are vulnerable, and experience greater mortality during early dry season fires than in the typically more intense late dry season fires (P.A. Werner, pers. comm. 1990). Thus it is hypothesized that it is these early dry season fires which cause greatest patchiness in the ground layer. On the other hand, at very low intensity or no fire, the existing plants grow in size, but little change in patchiness occurs. However, species may be gradually lost through competitive exclusion (Huston 1979), or richness increase due to absence of adversity or increase in favorableness, depending on the type of patch. Similarly, at high intensity, the ground cover receives a much more homogeneous treatment by fire, diminishing patchiness. Patchiness of the ground layer is thus also consistent with the intermediate disturbance hypothesis (Connell 1978).

The situation is different with the canopy. Typically, early dry season fires impact minimally on the canopy. In fact, the lack of impact on the flowering of fruit trees was a stated reason for the concentration of Aboriginal burning at this time (Haynes 1985). However, high-intensity late dry season fires typically scorch high into the canopy (Braithwaite and Estbergs 1985) and this occurs when the pre-wet season leaf flush is occurring and is likely to cause maximum mortality (Lonsdale and Braithwaite 1991).

At the no fire/low-intensity end, major change occurs due to the recruitment of plants out of the ground layer, as discussed in the section on woody and herbaceous plants above.

The pattern of change described in the preceding paragraphs is summarized in Fig. 7.8. Change in patchiness can be positive or negative depending on the loss or gain of individuals of sensitive plant species. The degree of change in relation to fire intensity in the canopy layer is the opposite to that for the ground layer. Fire shifts the advantage between the ground layer and the canopy layer, but in a fire-intensity-dependent way. If the intensity is low, dominance shifts towards the canopy and if it is high, it shifts towards the ground layer. At intermediate intensity, transition between the two is minimal. In combination the elements of this model can maintain savanna as a dynamic stasis. The key to whether the savanna changes to grassland or rain forest or remains as savanna is the ambient fire regime.

Denslow (1980) has argued that the most common patch size will support the highest diversity of species. The historic disturbance regime will produce patch sizes which have been most common historically. This is one reason why the documentation of the traditional fire regime of indigenous hunter-gatherers is of significance for contemporary conservation. However, as Braithwaite (1992) has argued, the traditional regime, as expressed on the landscape as a whole, is an emergent property of the activities of many quasi-independent individuals and groups. It was not the result of a brilliant ecological master-plan which we can simply take down from the historical shelf and naively apply in a contemporary context (Redford 1991) of roads, tourists, exotic invader problems, contemporary technology, and changed Aboriginal culture. It is necessary to reconstruct the historic pattern of burning as best we can using knowledge from any source.

Braithwaite (1991) described the seasonal pattern of Aboriginal burning using the historical record from the 19th century and found it corresponded well with the results of independent ethnographic studies. The traditional Aboriginal regime had a peak of burning in the early dry season. However, some fires did occur at any time during the 9.5-month fire season (Braithwaite 1991). It has been found by National Park managers that the more country that is burnt during the early dry season, the less that is burnt during the late dry season (Press 1988b). This is partly due to the fragmentation of the savanna fuel induced by the early fires. The subsequent late fires do not carry as far because of the lack of continuity of fuel. For the same reason, parcels of land remain unburnt in an area which had received heavy early burning. Thus historically, the three important elements necessary for the model to work were integral to the traditional Aboriginal burning regime, early fires with some late fires and some areas unburnt.

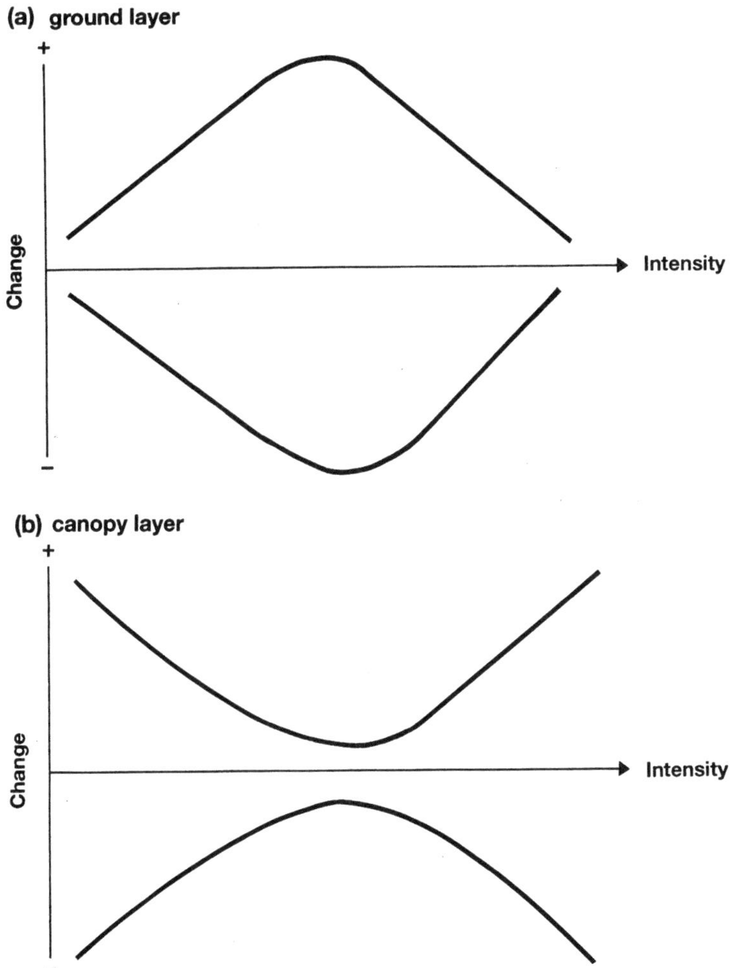

Fig. 7.8a.b. A model of change in biological attributes in relation to fire intensity in savannas. **a** Ground story. **b** Canopy story.

It follows from Denslow (1980) that the traditional regime described above would maximize species diversity. Thus the model would allow the persistence of the diversity of savanna vegetation. In contrast with the impoverished rain forest fauna which characterizes the cerrado of Brazil (Redford and da Fonseca 1986), the savanna of Australia has a distinctive and diverse mammal fauna richer than the local rain forest (Braithwaite et al. 1985). Thus it might be expected that at least the mammal fauna of the Australian savanna would also operate in synchrony with the patch dynamics of the savanna vegetation and there is some evidence that this happening (Fig. 7.9).

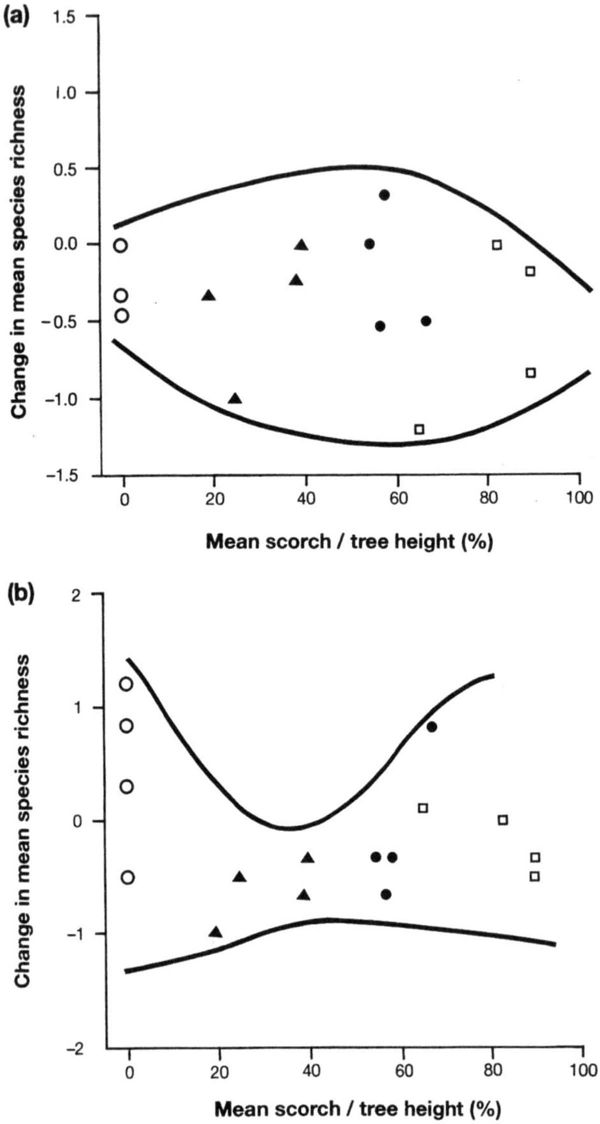

Fig. 7.9.a.b. Change in mean species richness of small mammals plotted against mean scorch/tree height (%) after 2 years experimental fires at Kapalga, Kakadu National Park, Australia. *Symbols* as for Fig. 7.7 **a** Terrestrial. **b** Arboreal. (R.W.Braithwaite, unpubl.)

7.7 Conclusion

In northern Australia, the present biodiversity of the savanna landscape appears to be maintained by an historic anthropogenic fire regime. It is probable that an approximation of the landscape pattern produced by the traditional Aboriginal burning regime maximizes biodiversity through maintaining habitat diversity, savanna patchiness, and species diversity, and protecting endemic species. The research task is to determine and optimize that fire mosaic.

Acknowledgments.
I thank my colleagues Alan Andersen, Garry Cook, Laurie Corbett, and Dick Williams for the use of their unpublished data, Mark Lonsdale, Philip Stott and Juan Silva for reading the manuscript, many colleagues for discussion, and WWF (Australia) for funding my mammal studies and the Australian Department of the Arts, Sport, Environment, and Territories for funding my trip to the workshop in Brasilia.

References

Andersen AN (1991) Responses of ground-foraging ant communities to three experimental fire regimes in a savanna forest of tropical Australia. Biotropica 23:575-85

Belsky AJ (1986) Population and community processes in a mosaic grassland in the Serengetti, Tanzania. J Ecol 74:841-856

Belsky AJ (1990) Tree/grass ratios in East African savannas: a comparison of existing models. J Biogeogr 17:483-490

Bowman DMJS (1991) Environmental determinants of *Allosyncarpia ternata* forests that are endemic to western Arnhen Land, northern Australia. Aust J Bot 39:575-89

Bowman DMJS, Fencham RJ (1991) Response of a monsoon forest-savanna boundary to fire protection, Weipa, northern Australia. Aust J Ecol 16:111-118

Bowman DMJS, Wightman GM (1985) Small scale vegetation pattern associated with a deeply incised gully, Gunn Point, northern Australia. Proc R Soc Queensl 96:63-73

Bowman DMJS, Wilson BA, Dunlop CR (1988a) Preliminary biogeographic analysis of the Northern Territory vascular flora. Aust J Bot 36:503-17

Bowman DMJS, Wilson BA, Hooper RJ (1988b) Response of *Eucalyptus* forest and woodland to four fire regimes at Munmarlary, Northern Territory, Australia. J Ecol 76:215-232

Bowman DMJS, Wilson BA, McDonough L (1991) Monsoon forests in northwestern Australia. I. Vegetation classification and the environmental control of tree species. J Biogeogr 18:679-686

Braithwaite RW (1985) The Kakadu fauna survey: an ecological survey of Kakadu National Park. Report to Australian National Parks and Wildlife Service, Canberra (unplub.)

Braithwaite RW (1987) Effects of fire regimes on lizards in the wet-dry tropics of Australia. J Trop Ecol 3:265-275

Braithwaite RW (1990) Australia's unique biota: implications for ecological processes. J Biogeogr 17:347-354

Braithwaite RW (1991) Aboriginal fire regimes of monsoonal Australia in the 19th century. Search 22:247-9

Braithwaite RW (1992) Black and green. J Biogeogr 19:113-116

Braithwaite RW, Estbergs JA (1985) Fire pattern and woody vegetation trends in the Alligator Rivers region of northern Australia. In: Tothill JC, Mott JJ (eds) Ecology and management of the worlds savannas. Aust Acad Sci, Canberra, pp 359-364

Braithwaite RW, Estbergs JA (1987) Firebirds of the Top End. Aust Nat Hist 22: 298-302

Braithwaite RW, Winter JW, Tatlor JA, Parker BS (1985) Patterns of diversity and structure of mammalian assemblages in the Australian tropics. Aust Mamm 8:171-86

Braithwaite RW, Cook GD, Williams RJ (1992) Towards an integrated study of sustainable land use in the Top End. Northwest Pastoral Conf Proc, DPIF, Katherine, pp 41-43

Connell JH (1978) Diversity in tropical rain forests and coral reefs. Science 199: 1302-1310

Cook GD (1991) Effects of fire regimen on two species of epiphytic orchids in tropical savannas of the Northern Territory. Aust J Ecol 16:537-540

Cook GD, Andrew MH (1991) The nutrient capital of indigenous *Sorghum* species and other understorey components of savannas in north-western Australia. Aust J Ecol 16:375-84

Crawford DN (1979) Effects of grass and fires on birds in the Darwin area. Emu 10:150-152

Denslow JS (1980) Patterns of plant species diversity during succession under different disturbance regimes. Oecologia (Berl) 46:18-21

Fencham RJ (1990) Interactive effects of fire frequency and site factors in tropical *Eucalyptus* forests. Aust J Ecol 15:255-266

Freeland WJ, Winter JW, Raskin S (1988) Australian rock-mammals: a phenomenon of the seasonally dry tropics. Biotropica 20:70-79

Frost P, Medina E, Menaut J, Menaut C, Solbrig O, Swift M, Walker B (eds) (1986) Responses of savannas to stress and disturbance: a proposal for a collaborative programme of research. Biol Int Spec Issue 10:1-82
Gillon D (1983) The fire problem in tropical savannas. In: Bourlière F (ed) Ecosystems of the World 13: Tropical savannas. Elsevier, Oxford, pp 617-641
Groombridge B (1992) Global biodiversity: status of the earth's living resources. A Report compiled by the World Conservation Monitoring Centre. Chapman & Hall, London
Haynes CD (1985) The pattern and ecology of *munwag:* traditional Aboriginal fire regimes in north-central Arnhemland. Proc Ecol Soc Aust 13:203-214
Hoare JRL, Hooper RJ, Cheney NP, Jakobsen KLS (1980) A report on the effects of fire in tall open forest and woodland with particular reference to fire management in Kakadu National Park in the Northern Territory. Report to the Australian National Parks and Wildlife Service, Canberra (unpubl)
Huston M (1979) A general hypothesis of species diversity. Am Nat 113:81-101
Jones R (1980) Hunters in the Australian coastal savanna, in human ecology. In: Harris DR (ed) Savanna environments. Academic Press, London, pp 107-146
Kellman M (1984) Synergistic relationships between fire and low soil fertility in neotropical savanna: a hypothesis. Biotropica 16:158-160
Lonsdale WM, Braithwaite RW (1991) Assessing the effects of fire on vegetation in tropical savannas. Aust J Ecol 16:363-74
Medina E, Huber O (1992) The role of biodiversity in the functioning of savanna ecosystems. In: Solbrig OT, van Emden HM, van Oordt PGWJ (eds) Biodiversity and global change. Union Biol Sci, Paris, Mongr 8:139-158
Mott JJ, Tothill JC, Weston EJ (1981) Animal production from the native woodlands and grasslands of northern Australia. J Aust Inst Agri Sci 47:132-141
Porter JW, Henderson R (1983) Birds and burning histories of open forest at Gundiah, south eastern Queensland. Sunbird 13:61-68
Press AJ (1988a) The distribution and status of macropods (Marsupialia: Macropididae) in Kakadu National Park, Northern Territory. Aust Mamm 11:103-108
Press AJ (1988b) Comparisons of the extent of fire in different land management systems in the Top End of the Northern Territory. Proc Ecol Soc Aust 15:167-175
Redford KH (1991) The ecologically noble savage. Cult Survival Q 15:46-48
Redford KH, da Fonseca GAB (1986) The role of gallery forests in the zoogeography of the Cerrado's non-volant mammalian fauna. Biotropica 18:126-135
Russell-Smith J (1991) Classification, species richness, and environmental relations of monsoon rain forest vegetation in the Northern Territory, Australia. J Veget Sci 2:259-278
Russell-Smith J, Lucas DE, Brock J, Bowman DMJS (1993). Allosyncarpia-dominated rain forest in monsoonal Australia. J Veget Sci 4:67-82
Solbrig OT (ed) (1991) Savanna modelling for global change. Biol Int Spec Issue 24:47. IUBS, Paris
Solbrig OT (1992) Biodiversity: an introduction. In: Solbrig OT, van Emden HM, van Oordt PGWJ (eds) Biodiversity and global change. Int Union Biol Sci, Paris, Monogr 8:13-20
Specht RL (1990) Climatic control of ecomorphological characters and species richness in mediterranean ecosystems of Australia. In: Specht RL (ed) Mediterranean type-ecosystems. Kluwer, Dordrecht, pp 149-155
Stott PA (1986) The spatial pattern of dry season fires in savanna forests of Thailand. J Biogeogr 13:345-358
Stott PA, Goldammer JG, Werner WL (1990) The role of fire in the tropical lowland deciduous forests of Asia. In: Goldammer JG (ed) Fire in the tropical biota: ecosystem processes and global challenges. Springer, Berlin Heidelberg New York, pp 32-44
Taylor JA, Dunlop CR (1985) Plant communities of the wet-dry tropics of Australia: the Alligator Rivers region, Northern Territory. Symp Ecol Soc Aust 13:83-127

Taylor JA, Tulloch D (1985) Rainfall in the wet-dry tropics : extreme events at Darwin and similarities between years during the period 1870-1983. Aust J Ecol 10:281-95
Trainor CR, Woinarski JCZ (1994) Response of lizards to three experimental fire regimes in the savanna forests of Kakadu National Park. Wildl Res 21:131-48
Walker BH, Nix H (1993) Managing Australia's biological diversity. Search 24:173-8
Walter H (1971) Ecology of tropical and sub-tropical vegetation. Oliver & Boyd, Edinburgh
Woinarski JCZ (1990) Effects of fire on the bird communities of tropical woodlands and open forests in northern Australia. Aust J Ecol 15:1-22
Woinarski JCZ, Braithwaite RW (1990) The terrestrial vertebrate fauna and vegetation of the Kakadu Conservation Zone. Report to Resource Assessment Commission, Canberra (unpubl)

8 Diversity of Herbivorous Insects and Ecosystem Processes

8 Diversity of Herbivorous Insects and Ecosystem Processes

Thomas M. Lewinsohn and Peter W. Price

8.1 Introduction

Most general accounts of savanna ecology give little consideration to phytophagous animals (Walter 1971; Schnell 1973; Goodland and Ferri 1979; Sarmiento 1984). The major exception, of course, are African ungulate herbivores, whose impact on savanna and grassland ecosystems has received some attention (McNaughton 1976; Sinclair 1983). Plant-eating insects, on the other hand, are hardly mentioned except as pollinators (Cole 1986).

For savannas, as for most tropical and subtropical ecosystems, there are virtually no estimates of insect diversity. Nonetheless, insect contribution to total species diversity and consumer biomass is orders of magnitude greater than that of vertebrates (Lamotte 1975; Gillon 1983). There is no reliable breakdown of this diversity into lower categories but, although to a large extent arthropod biomass is concentrated in predators (e.g., spiders, ants) and detritivores (e.g., termites), phytophages probably account for a large share of species diversity, and possibly biomass as well.

In savannas as in most other terrestrial ecosystems we thus meet with an initial predicament: are phytophagous insects being neglected out of ignorance of their roles or are they disproportionately less important to ecosystem processes than their high diversity would indicate?

Food-web structure adds to this predicament. We are not aware that any savanna food web has ever been comprehensively studied. An approximation may be found in Polis (1991), whose synthesis of extensive studies in a warm temperate desert in California reports an estimated 2000 - 3000 insect species.

Ecological Studies, Vol. 121
Solbrig, Medina and Silva (eds) Biodiversity and Savanna Ecosystem Processes

These include, at a guess, a minimum of 1000 phytophages feeding on 174 species of vascular plants. Even after drastic simplification, the total food web of 30 "kinds of organisms" still includes 22 220 food chains. For food links, as much as for individual species, it is hardly conceivable that every single one will perform a unique, though not necessarily essential, role within ecosystem processes. How many of these species and trophic links are redundant (Lawton and Brown 1993) is unknown.

We are not dealing with the question whether and how herbivores and particularly insect herbivores, taken together, affect ecosystem processes; though even that cannot be taken for granted (see for instance the "green Earth" polemic, Hairston et al. 1960, critiques and rejoinders in Hazen 1970). We are asking whether plant-eating insect diversity has verifiable effects on ecosystem processes.

Given our overwhelming ignorance of even the number of species in savannas – except for some groups of herbivorous insects in a very few sites – this may seem a foredoomed task. Nevertheless, given that total species counts, even in comparative surveys, will not by themselves provide sufficient answers, we should instead look for an approach that can lead to more restricted and answerable questions.

In this chapter first we discuss in general terms how insect diversity could be related to ecosystem processes. Second, we consider functional categories of herbivorous insects suitable for the empirical investigation of links between herbivore diversity and ecosystem processes. We make no attempt at presenting a review of the relevant information, which is widely scattered, fragmented, and mostly inferential for our purposes. Given the scarcity of data for savannas, this discussion is bound to be very general, though we attempt to emphasize features that are particularly relevant to these ecosystems.

8.2 What Links?

Insect herbivore diversity could in principle be linked to ecosystem processes either directly or indirectly by way of its effects on plant diversity (Fig. 8.1). Few studies have attempted to link total insect and plant diversity, and these are especially concerned with temperate grasslands (Murdoch et al. 1972; Southwood et al. 1979). Such studies are mostly correlative but, explicitly or not, a greater influence of plant diversity on insects is assumed than the reverse. Likewise, much more work has been done on plant community responses to ecosystem processes than in the other direction.

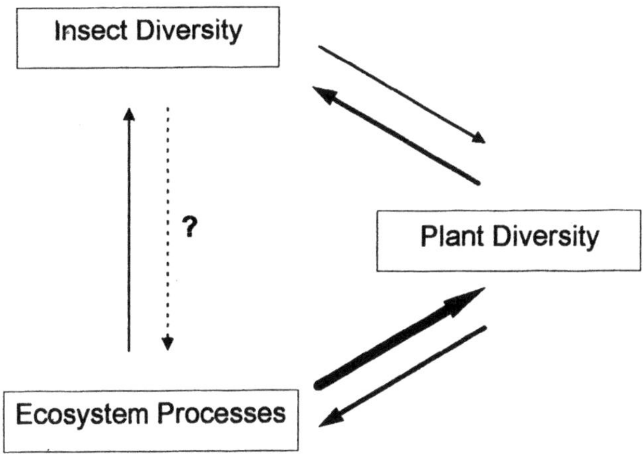

Fig. 8.1. Insect herbivore diversity can influence ecosystem processes either directly or through its effects on plant diversity. *Arrow widths* indicate the relative amount of information available on each link; for example, more work has been done on the responses of plant diversity to ecosystem processes than the converse

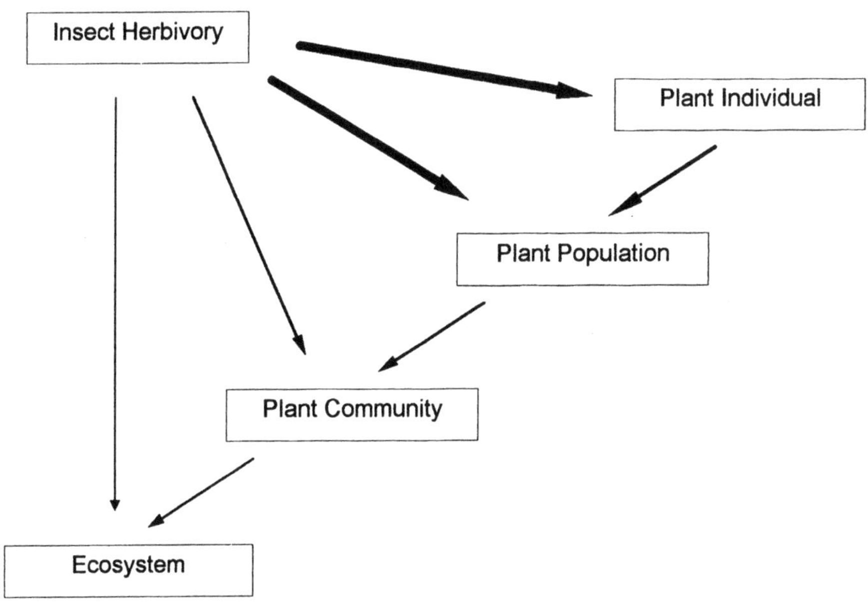

Fig. 8.2. A classical hierarchy of organizational levels from individuals to the ecosystem. Theoretically, herbivory can affect any level directly or indirectly through preceding ones. *Arrow widths* give a rough indication of the amount of information on each link from existing studies. They do not imply actual importance of each link

For our present aim, however, most available studies are of little use because they provide information on one or the other link but rarely on both (Fig. 8.1). Ecosystem effects on plants cannot be automatically assumed to "follow through" to insect communities. In the other direction, with which we are concerned here, our understanding is even more tenuous.

Studies of insect-plant interactions have mostly focused on lower organizational levels (Fig. 8.2), that is to say, the effects of insects (populations in general) on plant individuals and populations, and vice versa (Crawley 1983, 1989; Huntly 1991). Investigations of the effects on entire plant communities are much scarcer and even more so on ecosystems (Strong et al. 1984; Huntly 1991). Effects of herbivorous insect communities (or assemblages, to use a more neutral term) on plants are again less studied, and almost entirely restricted to populations of one or a very few host species.

Direct effects of herbivore populations and assemblages on ecosystem properties are readily conceivable and some instances will be commented below. However, we contend that the majority of such effects are indirect. Therefore, any endeavor to explain them should be more fruitful on the lines of the mechanistic protocol outlined in Fig. 8.2. In other words, to recognize if and how changes in insect herbivore diversity affect ecosystem properties, we need to ask how they affect plant individuals, populations, and communities in turn.

8.3 Functional Categories

Phytophagous insect surveys are often classed solely in taxonomic groupings (Gillon 1983). Other, more functional classifications divide insects in "guilds" based on their feeding mode, such as chewers, borers, and suckers, which may be further subdivided into, for instance, the pit-feeding and strip-feeding folivores (Root 1973). Moreover, such feeding modes may also be cross-classified by the plant organs on which they feed (Lawton 1982; Lawton et al. 1993).

The functional divisions we present in Fig. 8.3 focus on potential consequences of phytophage activity rather than their taxonomic status or feeding mode. This requires additional explanation.

Fig. 8.3. (Page 147) A simplified outline of functional categories of herbivorous insects (*left column*) and their potential effects on various levels of organization (*other columns*). Feedback effects are left out for simplicity. The *major boxes in the top row* comprise general effects for each level and are linked to all major effects in adjacent levels. Specific links deemed especially important are suggested by the remaining *arrows*. Additional explanations in text

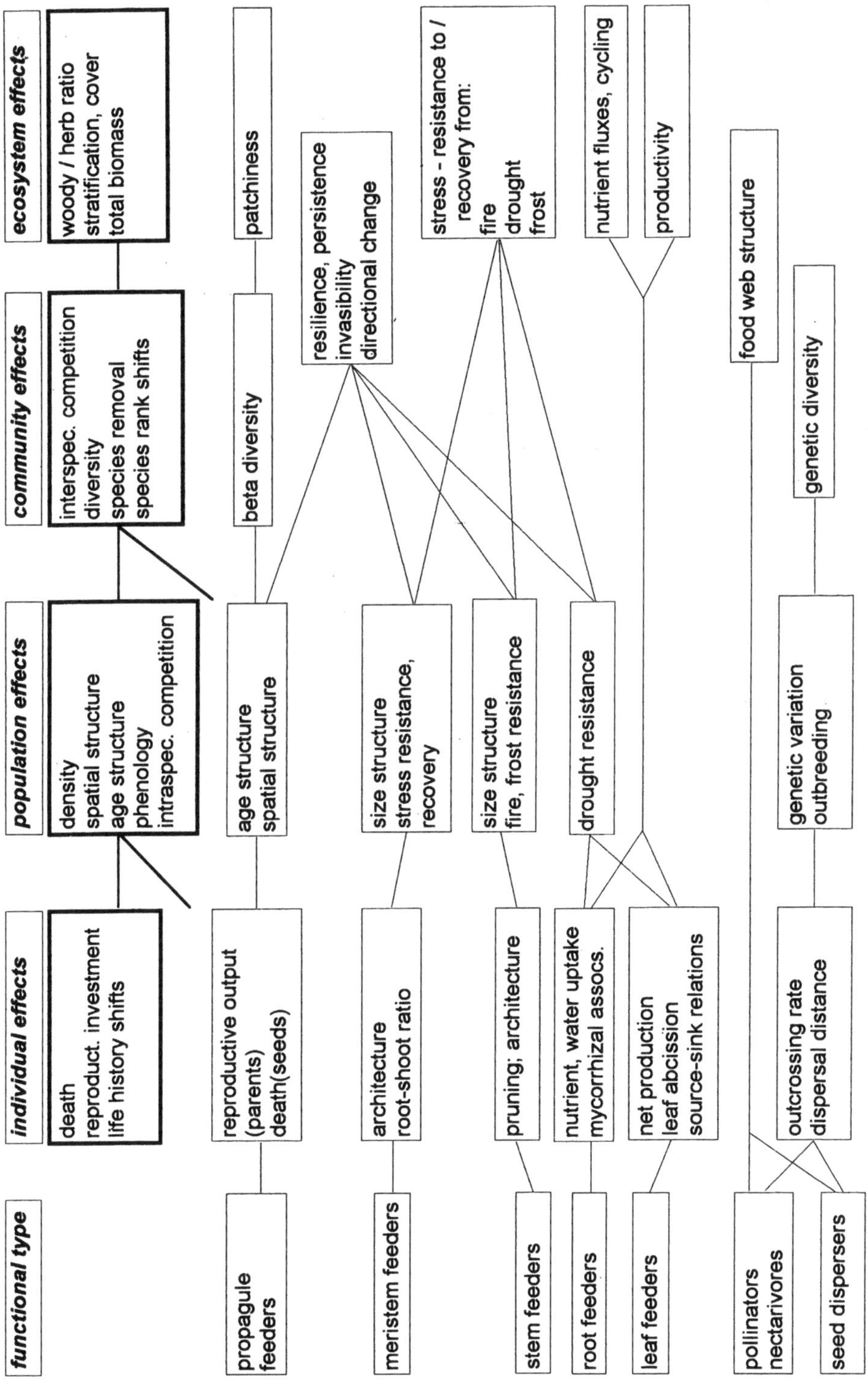
functional type
individual effects
population effects
community effects
ecosystem effects
death
reproduct. investment
life history shifts
density
spatial structure
age structure
phenology
intraspec. competition
interspec. competition
diversity
species removal
species rank shifts
woody / herb ratio
stratification, cover
total biomass
propagule feeders
reproductive output (parents)
death(seeds)
age structure
spatial structure
beta diversity
patchiness
meristem feeders
architecture
root-shoot ratio
size structure
stress resistance,
recovery
resilience, persistence
invasibility
directional change
stem feeders
pruning; architecture
size structure
fire, frost resistance
stress - resistance to / recovery from:
fire
drought
frost
root feeders
nutrient, water uptake
mycorrhizal assocs.
drought resistance
nutrient fluxes, cycling
leaf feeders
net production
leaf abcission
source-sink relations
productivity
pollinators
nectarivores
outcrossing rate
dispersal distance
genetic variation
outbreeding
genetic diversity
food web structure
seed dispersers

We distinguish mutualistic phytophages (pollinators, nectarivores, seed dispersers) from antagonists (all other groups), though recognizing that consumption by supposed mutualists is not necessarily matched by their services, and that certain "antagonists" conceivably may benefit plants at some level (Owen and Wiegert 1987).

Another major contrast we emphasize is that of propagule and meristem feeders (plus the mutualistic groups) versus stem, root, and leaf feeders. The effects of the first group are disproportional to the actual biomass they consume, whereas in the second group biomass consumption should be more directly related to the intensity of their effects on plants.

Though they are referred to plant parts, these types cannot be strictly separated by plant morphology alone. Meristems, for instance, are not autonomous but belong to plant parts such as branches or roots. Although many insects (e.g., moth caterpillars, weevil larvae) are indeed specialized meristem feeders, many others will consume developed organs together with buds.

Propagule feeders, according to effects on sexual reproduction and propagation, should encompass fruit and seed feeders as well as flower-bud and flower feeders (but not legitimate flower visitors and seed dispersers, though these consume pollen, nectar, or seed rewards).

Insect feeding modes are spread across these functional categories. For instance, the term "leaf feeders" includes leaf chewers, miners, gall-makers, and suckers. The emphasis, as we said before, is on their effects on plants, both immediate and ultimate: thus, leaf feeders are primarily reducing photosynthetic area or efficiency, or else consuming plant production straight from the source.

Sucking insects might be separated into their own category of sap feeders, considering that their effect is less dependent on feeding site than on gross consumption of circulating photosynthate (phloem feeders) or inorganic nutrients (xylem feeders).

Galls have different kinds of effects on host plants. They often act as nutrient or photosynthate sinks, and so gall-makers could be grouped with sap-suckers (Moran and Whitham 1990). Beyond this, some gallers, notably stem and branch gallers, can also modify plant architecture by changing growth and branching patterns, in common with meristem, stem, and root feeders.

8.4 Linking Herbivory to Effects on Plants and Ecosystems

In attempting to establish a comprehensive model we have to organize a multiplicity of connections between and within levels. Rather than smother Fig. 8.3 in arrows connecting everything with everything, we have grouped some key processes in the first row of boxes in the figure. These should be read as being connected to virtually every component in neighboring levels, and to each other as well. We also omit feedbacks to herbivore groups for readability, but they are certainly there.

Particular links between other components of Fig. 8.3 reflect an evaluation of their importance rather than exclusiveness. Indeed, it is difficult to identify any path which could be considered to be autonomous from all other compartments and connections. As we move up the hierarchy from individuals to ecosystems across Fig. 8.3, links become more difficult to trace. This means that organizing testable hypotheses on comprehensive paths, from herbivores up to particular ecosystem processes, is also that much harder.

Intergradation of community and ecosystem effects is represented in Fig. 8.3 by boxes placed midway between their columns. Insofar as several ecosystem properties are described by their biotic components and refer particularly to species composition and relations, they are indistinguishable from community characteristics.

8.5 Herbivore Effects on Plant Individuals and Populations

Direct mortality from insect herbivory is relatively uncommon. Nonetheless, herbivory will often increase mortality by making plants more susceptible to other herbivores, parasites, diseases, and stress such as subsequent frost damage (Crawley 1983, 1989; Strauss 1991).

Many effects of herbivores on plants are subtle though verifiable. For instance, leaf galls act as metabolic sinks and may induce premature leaf abscission (Faeth et al. 1981; Williams and Whitham 1986). A variety of such effects have been extensively investigated (reviews in Harper 1977; Crawley 1983, 1989), and some are exemplified in Fig.8.3.

Two kinds of effects are worth emphasizing. First, insect outbreaks can limit reproduction drastically. In cerrados, seed crops are often almost entirely destroyed by predispersal predation (Lewinsohn, pers. observ.). Effective recruitment from seeds can thus be limited to particularly favorable years and result in pulsed and unstable population age structures; also, in a shift to vegetative reproduction. At the community level these contribute to heterogeneity and increase spatial diversity, including the beta (intersite) species diversity component; such processes can be enhanced by herbivory if it is spatially structured, as for instance in the distance-dependent Janzen-Connell model (Janzen 1970).

The second set of responses we emphasize refers to changes in plant architecture due to pruning or meristem destruction (Fig. 8.3). Their manifold consequences include: changes in intra- and interspecific competitive ability; phenological or life-history shifts such as delayed flowering; and changes in susceptibility to fire and frost damage. The latter are of special interest in savannas and may affect the imbalance between woody and herbaceous vegetation (see below). Plant morphology, at least of herbaceous species, may also be changed by leaf-eating and sap-sucking insects (Meyer 1993).

Individual or population compensation can deflect the consequences of apparent herbivore damage. In particular, effects on individual reproduction cannot be inferred from observed losses to herbivory (Whitham et al. 1991). Ant protection of *Cassia fasciculata* reduced leaf area loss to herbivores and increased growth rate, but this did not increase seed set (Kelly 1986). Plant responses can also be largely modified by additional factors. Plants on more fertile soils are supposedly better able to compensate for loss to herbivores; yet experimental results show the contrary for *Solidago altissima* (Meyer and Root 1993).

Damage can also be compensated through demographic responses. An experimental study of two native *Haplopappus* species in California showed that, whereas both suffered appreciable seed loss, herbivore exclusion with insecticides increased seedling recruitment in only one of them; seed predation had no effect on the population dynamics of the other (Louda 1983). Thus, another potential fallacy lies in assuming that every reduction in plant performance leads to a population decrease (Crawley 1989).

A large extent of our knowledge of the effect of herbivorous insects on plant populations is derived from introductions of insects to control weeds (Julien 1987). Along with many failures, biological control programs include a number of successful cases that show unambiguously the capacity of insects to provoke substantial reduction and even local extinction of particular host plants. Nevertheless, they are of restricted help in understanding natural ecosystems because they are biased towards specialized herbivores released in large numbers and their target populations are usually aliens in cultivated or highly disturbed systems (Crawley 1989).

8.6 Herbivore Effects on Communities

We expect herbivory to affect plant communities most often by altering relationships among plant species. An intense impact on a numerically or functionally dominant (or keystone) plant could diminish its rank and not only allow other species to increase, but even raise plant species richness and diversity. An equivalent impact on a rare plant will obviously not translate into community effects. Such shifts in plant competitive efficiency have been documented a number of times (Crawley 1989; Huntly 1991). Three chrysomelid species feeding preferentially on *Solidago canadensis* kept it at less than 1% cover in old fields; when herbivores were excluded with insecticides, the plant expanded to 40 to 70% cover at the expense of several early successional species. In this case succession is influenced by herbivory, though no species were excluded or added in either case (McBrien et al. 1983).

Food-web structure is at least as important as species numbers or rank-abundances for community dynamics. The total pattern of connections of herbivore and plant species comprises both the host specialization of herbivores and the size and overlap of species assemblages associated with different plants. A polyphagous herbivore with frequency-dependent host choice, for instance, would tend to switch to more abundant species and thus help to regulate them, with relatively lower pressure on rarer species. However, a rare plant may be driven to local extinction by one or several herbivores shared with more common plants. This has been shown for a composite forb excluded locally by a grasshopper that feeds on another composite (Parker and Root 1981; see also Futuyma and Wasserman 1980; Thomas 1986).

As we stated before, effects on communities intergrade with ecosystem effects. Foremost instances are species diversity and its components, and food-web structure (Fig. 8.3) which we include under both effects.

8.7 Direct Effects of Herbivores on Ecosystems

The effects of herbivore outbreaks on nutrient cycling and fluxes have received some attention. Folivores could accelerate nutrient release from living plants; so could sap-suckers which release large quantities of carbohydrate-rich honeydew.

Defoliation by the spring cankerworm (Lepidoptera: Geometridae) in outbreak years consumed up to 33% leaf mass in a mixed hardwood forest and was accompanied by up to fivefold increases of stream-exported nitrates (Swank et al. 1981). However, with similarly intensive outbreaks in the Hubbard Brook experiment, where baseline concentrations are higher and very variable, no such increase was detected (Bormann and Likens 1979). Other effects of massive cankerworm defoliation included reduction of wood production and increases in foliage production, frass deposition, leaf litterfall, litter metabolism, and mineral nutrient availability (Swank et al. 1981).

The effects of aphid honeydew in an *Alnus rubra* plantation, contrary to previous hypotheses, did not enhance bacterial activity and soil nitrogen. On the contrary, aphids caused significant reductions of available soil nitrogen (as proposed by Owen and Wiegert 1987), of tree nitrogen uptake and of net primary production - reducing wood and bark but not foliage production (Grier and Vogt 1990).

Very few other ecosystem properties have been directly related to particular insects. The previously mentioned study of early sucessional species enhancement due to herbivory on a dominant plant is a case in point (McBrien et al. 1983). Although Mattson and Addy (1975) postulated a long-term cybernetic role for herbivory as regulator of ecosystem primary productivity, this has found little support (Swank et al. 1981).

8.8 Indirect and Nonadditive Effects of Herbivore Assemblages

Virtually all data on herbivore-plant effects at whatever level concern a particular herbivore species. To understand the effects of herbivore diversity we also need information on how herbivorous species interact and if interactions alter effects on plants.

Two kinds of questions come to mind: first, is there significant interspecific competition among herbivores on the same host plant? Second, are effects of herbivores on plants additive, or are there indirect effects which cannot be predicted from single herbivore-plant effects? Although the questions are interconnected, the first has been of especial interest to community ecologists in the 1960s and 1970s, with less regard to effects on plants (Strong et al. 1984). As for the second, evidence from experimental studies is scarcer and more recent.

In an experimental study of *Rhus glabra* (Anacardiaceae), deer and two herbivorous beetles, a bud- and leaf-eating chrysomelid and a stem and rhizome-boring cerambycid, Strauss (1991) found several kinds of interaction, mostly sequential. For instance, cerambycid attack was more damaging to plants in combination with chrysomelid feeding and more likely on shoots previously attacked by chrysomelids. Thus, the history of past herbivory as well as current interactions generates significant nonadditive effects of herbivores on plants (see also Karban and Strauss 1993).

Classical competition theory predicts strongest interactions among species that exploit a common set of resources with similar modes: within guilds, that is. Although a lot of effort has been spent on interactions of similar herbivorous species on a common set of plants (Strong et al. 1984), recent work indicates that in some cases species with distinct feeding habits may interact strongly though indirectly, by way of the host plants. For instance, root- and leaf-feeding insects show measurable effects on each other (Moran and Whitham 1990) and, conversely, they have nonadditive effects on host plants (Masters et al. 1993).

Biological control programs with concomitant introduction of several insects provide additional evidence that interactions among herbivores are essential to their outcome or, in other words, that effects on target plants are nonadditive (for a successful case of weed control by three herbivore species, see McEvoy et al. 1991).

8.9 Proximal and Ultimate Effects on Ecosystems

The ecosystem features in Fig. 8.3 range from directly measurable ones to ultimate properties. The first include landscape-level physionomical descriptors, on the one hand, for example overall cover and stratification, as well as "patchiness"; on the other, functional properties such as nutrient fluxes and overall biomass and productivity.

At the other extreme, there are metaproperties such as overall system stability and resilience which can be ascertained only from the ecosystem's behavior over extended time periods and in response to all forms and levels of perturbation. These properties are very hard to establish both for practical reasons and because they are highly theory- and definition-dependent.

Halfway between these extremes are properties which describe ecosystem responses to particular perturbations. Resistance and recovery from various forms of stress are part of the ecosystem's overall resilience; however, they are more amenable to investigation and of direct interest to researchers and management given the kind of stress to which a system is prone. The ones we emphasize - fire, drought, and frost - are known to be important in various kinds of savannas, and potentially the best candidates for a comprehensive research program; they are also readily related to other features.

In savannas, the woody-herbaceous ratio and the dicot-grass ratio, are known to be especially important to ecosystem function, properties, and changes (see Chaps. 1 and 2). Due to major differences in morphology, physiology, and life-form, these plant groups support distinctive sets of herbivores and this can be explored advantageously to investigate effects of various modes of herbivory on these imbalances and their attendant consequences.

8.10 Some Suggestions

It is very unlikely that usable data for the questions we address can be gleaned from studies carried out for other purposes. We cannot expect much progress from correlational studies or broad literature surveys. Instead, if these questions are deemed sufficiently important, they must be addressed by research programs of their own. Field studies should be designed to evaluate herbivore effects on plant growth, reproduction, dispersal, and recruitment, and not simply "to determine whether communities differ when herbivores are present or absent" (Huntly 1991).

Integration of population and community ecology with ecosystem studies is beset by some fundamental problems. Not surprisingly, it appears to be much easier to detect responses of populations or communities to ecosystem properties. Ecosystem-level measurements are relatively coarse, whereas effects of particular species or species sets are liable to be subtle (Vitousek 1990). We thus expect species biomass, and their biomass consumption rates as well, to be rather vague indicators of their effects on community and ecosystem behavior. For instance, the removal of some scarce and inconspicuous herbivores with key links to plants or other organisms could have stronger effects than reducing or eliminating much more abundant species. Specialized pollinators and dispersers are obvious instances of this, but there are certainly many other equivalent roles in ecosystems.

We also have to ask whether insect herbivores should be investigated as a separate entity from other organisms. Not only are they subject to parasitoids and predators but, more importantly, there are numerous indications that their effects on plants may be considerably modified by interactions with other taxa: vertebrate grazers (Danell and Huss-Danell 1985, Whitham et al. 1991), fungi and bacteria (pathogenic, decomposing, mycorrhyzae and so on (Clay et al. 1993). Thus, to understand fully the consequences of various combinations of herbivores on ecosystem processes, it may be necessary to take interactions with other relevant organisms into account as well.

To investigate the potential effects of herbivore diversity on ecosystem processes in savannas one should perhaps concentrate on a single, or possibly a few, plant species known to have key effects on a distinctive and measurable ecosystem feature. Manipulating a small set of the more important herbivores, singly and in combination, should allow testing hypotheses on their effects. Although this mechanistic approach is bound to be laborious and time-consuming – and more so in savannas than in simpler grasslands – it should produce much better answers than comparative and correlative studies or than experiments on single herbivore-plant pairs.

Acknowledgments.
TML received support from IBAMA and SCOPE to attend the Brasília meeting. PWP received support from NSF grants BSR-9020317 and DEB-9318188.

References

Bormann FH, Likens GE (1979) Pattern and process in a forested ecosystem. Springer, Berlin Heidelberg New York

Clay K, Marks S, Cheplick GP (1993) Effects of insect herbivory and fungal endophyte infection on competitive interactions among grasses. Ecology 74:1767-1777

Cole MM (1986) The savannas: biogeography and geobotany. Academic Press, London

Crawley MJ (1983) Herbivory; the dynamics of animal-plant interactions. Blackwell, Oxford

Crawley MJ (1989) Insect herbivores and plant population dynamics. Annu Rev Entomol 34:531-564

Danell K, Huss-Danell K (1985) Feeding by insects and hares on birches affected by earlier moose browsing. Oikos 44:75-81

Faeth SH, Connor EF, Simberloff D (1981) Early leaf abcission: a neglected source of mortality for folivores. Am Nat 117:409-415

Futuyma DJ, Wasserman SS (1980) Resource concentration and herbivory in oak forests. Science 210:920-922

Gillon Y (1983) The invertebrates of the grass layer. In: Bourlière F (ed) Tropical savannas. Elsevier, Amsterdam, pp 289-311

Goodland R, Ferri MG (1979) Ecologia do Cerrado. Itatiaia and Editora da Univ São Paulo, Belo Horizonte and São Paulo

Grier CC, Vogt DJ (1990) Effects of aphid honeydew on soil nitrogen availability and net primary production in an *Alnus rubra* plantation in western Washington. Oikos 57:114-118

Hairston NG, Smith FE, Slobodkin LB (1960) Community structure, population control, and competition. Am Nat 94:421-425

Harper JL (1977) Population biology of plants. Academic Press, London

Hazen WE (ed) (1970) Readings in population and community structure, 2nd edn. Saunders, Philadelphia

Huntly N (1991) Herbivores and the dynamics of communities and ecosystems. Annu Rev Ecol Syst 22:477-503

Janzen DH (1970) Herbivores and the number of tree species in tropical forests. Am Nat 104:501-528

Julien MH (ed) (1987) Biological control of weeds: a world catalogue of agents and their target weeds, 2nd edn. CAB International, Wallingford

Karban R, Strauss SY (1993) Effects of herbivores on growth and reproduction of their perennial host, *Erigeron glaucus*. Ecology 74:39-46

Kelly CA (1986) Extrafloral nectaries: ants, herbivores and fecundity in *Cassia fasciculata*. Oecologia 69:600-605

Lamotte M (1975) The structure and function of a tropical savannah ecosystem. In: Golley FB, Medina E (eds) Tropical ecological systems. Springer, Berlin Heidelberg New York, pp 179-222

Lawton JH (1982) Vacant niches and unsaturated communities: a comparison of bracken herbivores at sites on two continents. J Anim Ecol 51:573-595

Lawton JH, Brown VK (1993) Redundancy in ecosystems. In: Schulze ED, Mooney HA (eds) Biodiversity and ecosystem function. Springer, Berlin Heidelberg New York, pp 265-270

Lawton JH, Lewinsohn TM, Compton SG (1993) Patterns of diversity for the insect herbivores on bracken. In: Ricklefs RE, Schluter D (eds) Species diversity in ecological communities. Univ Chicago Press, Chicago, pp 178-184

Louda SM (1983) Seed predation and seedling mortality in the recruitment of a shrub, *Haplopappus venetus* (Asteraceae), along a climatic gradient. Ecology 64:511-521

Masters GJ, Brown VK, Gange AC (1993) Plant-mediated interactions between above- and below-ground insect herbivores. Oikos 66:148-151

Mattson WJ, Addy ND (1975) Phytophagous insects as regulators of forest primary productivity. Science 190:515-522

McBrien H, Harmsen R, Crowder A (1983) A case of insect grazing affecting plant succession. Ecology 64:1035-1039
McEvoy P, Cox C, Coombs E (1991) Successful biological control of ragwort, *Senecio jacobaea*, by introduced insects in Oregon. Ecol Applic 1:430-442
McNaughton SJ (1976) Serengeti migratory wildebeest: facilitation of energy flow by grazing. Science 191:92-94
Meyer GA (1993) A comparison of the impacts of leaf- and sap-feeding insects on growth and allocation of goldenrod. Ecology 74:1101-1116
Meyer GA, Root RB (1993) Effects of herbivorous insects and soil fertility on reproduction of goldenrod. Ecology 74:1117-1128
Moran NA, Whitham TG (1990) Interspecific competition between root-feeding and leaf-galling aphids mediated by host plant resistance. Ecology 71:1050-1058
Murdoch WW, Evans FC, Peterson CH (1972) Diversity and pattern in plants and insects. Ecology 53:819-829
Owen DF, Wiegert RG (1987) Leaf eating as mutualism. In: Barbosa P, Schultz JC (eds) Insect outbreaks. Academic Press, San Diego, pp 81-95
Parker MA, Root RB (1981) Insect herbivores limit habitat distribution of a native composite, *Machaeranthera canescens*. Ecology 62:1390-1392
Polis GA (1991) Complex trophic interaction in deserts: an empirical critique of food-web theory. Am Nat 138:123-155
Root RB (1973) Organization of a plant-arthropod association in simple and diverse habitats: the fauna of collards (*Brassica oleracea*). Ecol Monogr 43:95-124
Sarmiento G (1984) The ecology of neotropical savannas. Harvard Univ Press, Cambridge (transl by Otto Solbrig)
Schnell R (1973) Introduction à la phytogéographie des pays tropicaux. 2. Les milieux -- les groupements végétaux. Gauthier-Villars, Paris
Sinclair ARE (1983) The adaptations of African ungulates and their effects on community function. In: Bourlière F (ed) Tropical savannas. Elsevier, Amsterdam, pp 401-426
Southwood TRE, Brown VK, Reader PM (1979) The relationship of plant and insect diversities in succession. Biol J Linn Soc 12:327-348
Strauss SY (1991) Direct, indirect, and cumulative effects of three native herbivores on a shared host plant. Ecology 72:543-558
Strong DR, Lawton JH, Southwood R (1984) Insects on plants: community patterns and mechanisms. Blackwell, Oxford
Swank WT, Waide JB, Crossley DA Jr, Todd RL (1981) Insect defoliation enhances nitrate export from forest ecosystems. Oecologia 51: 297-299
Thomas CD (1986) Butterfly larvae decrease host plant survival in the vicinity of alternate host species. Oecologia 70:113-117
Vitousek PM (1990) Biological invasions and ecosystem processes: towards an integration of population biology and ecosystem studies. Oikos 57:7-13
Walter H (1971) Ecology of tropical and subtropical vegetation. Oliver and Boyd, Edinburgh
Whitham TG, Maschinski J, Larson KC, Paige KN (1991) Plant responses to herbivory: the continuum from negative to positive and underlying physiological mechanisms. In: Price PW, Lewinsohn TM, Fernandes GW, Benson WW (eds) Plant-animal interactions: evolutionary ecology in tropical and temperate regions. Wiley, New York, pp 227-256
Williams AG, Whitham TG (1986) Premature leaf abscission: an induced plant defense against gall aphids. Ecology 67:1619-1627

9 Biodiversity and Stability in Tropical Savannas

9 Biodiversity and Stability in Tropical Savannas

Juan F. Silva

9.1 Introduction

The stability of savanna plant communities depends on the kind, intensity, and extent of perturbation. This chapter will discuss some aspects of the nature of perturbations taking place in savanna communities, and will put forward the hypothesis that the relative stability of a community on the face of a perturbation depends on its floristic structure. The structure of a plant community, or floristic structure, is used here in a phytosociological sense, involving not only the species composition but also the relative importance of each species. This hypothesis implies that two savanna communities exhibiting the same values of diversity (namely the same diversity index H') may differ in stability if they differ in floristic structure. The hypothesis is based on the fact that individuals from different species react differently to changes in their environment. In turn, this dissimilarity in stability will also depend on the nature of the perturbation.

Individual plant responses will affect survival and fertility at the population level. The magnitude of the connection from individual to population will depend on the species' life history traits. This is another reason why community responses to perturbations depend heavily on the floristic structure.

A perturbation can be defined as a sudden and relatively short change in the prevalent environmental conditions. Some common perturbations in savannas are changes in the prevalent climate or fire regime and in the patterns of human use (such as cattle grazing or agriculture). These disturbances have two kinds of effects: (1) those that directly modify community structure (by differential removal of individuals as in grazing or by total

Ecological Studies, Vol. 121
Solbrig, Medina and Silva (eds) Biodiversity and Savanna Ecosystem Processes

replacement as in agriculture); (2) indirect effects that modify some of the prevailing physical conditions such as plant available moisture (PAM), plant available nutrients (PAN) and fire regime. As discussed above, these indirect effects may also result in variations in the structure of the community. In turn, changes in the relative importance of some species may affect the status of dependent or competing species, generating further modifications in community structure.

Stability and related properties of ecosystems are important concepts that have been formulated in different ways, using very variable terms and definitions (Harrison 1979; Connell & Sousa 1983; Holling 1986). This chapter will use stability as the property enabling a community to maintain its floristic structure on the face of an environmental perturbation. Although modeling of savanna stability has already been undertaken (Walker et al. 1981; Walker and Noy-Meir 1982), the relationships between stability and diversity are little known in tropical savannas.

9.2 Morphological and Functional Differences in Savanna Species

Different responses are expected from species differing widely in life-form, as in the case of grasses vs. savanna trees. However, species sharing the same life form, e.g., grasses, may also respond differently depending on various functional traits like photosynthetic metabolism, biomass allocation, phenology, etc. That so many grass species coexist in the same savanna is possible precisely because they differ in many traits (Sarmiento and Monasterio 1983). Coexisting savanna grasses differ in their life cycles, phenology, architecture, annual seed production, seed dormancy and germination, and photosynthetic metabolism (Sarmiento 1983a; Sarmiento and Monasterio 1983; Mott and Andrew 1985; Silva and Ataroff 1985; Silva 1987; Raventós and Silva 1988; Ernst et al. 1991; Veenendaal et al. 1993). Although the mechanisms are still to be studied, these differences result in contrasting responses to water stress, burning, shading, competition, and grazing (O'Connor 1993; O'Connor and Pickett 1992; Raventós and Silva 1988, 1995; Silva et al. 1990).

Tree species also show a wide array of morphological and functional differences, although they are not as well documented as for grasses. In neotropical savannas, tree species vary along a continuum from evergreen, sclerophyllous, and fire-resistant species at one extreme to deciduous, mesophyllous, and non-fire-resistant ones at the other. Although evergreen trees have been grouped together because they are reproductively active during the dry season (Sarmiento and Monasterio 1983), more detailed comparisons show that even closely related species differ in their reproductive phenologies and in other aspects of their reproductive biology (Oliveira and Silva 1993). Although almost half of the Brazilian savanna (*cerrado*) tree species are dispersed by wind, there is a wide variation in dispersal syndromes and type of propagules (Oliveira and Moreira 1992).

Tree leaves vary from large, very sclerophyllous (such as in *Palicourea rigida*) to small, coriaceous leaves (such as in *Casearia sylvestris*), and these differences may be related to their responses to water stress and to nutrient stress (Goldstein and Sarmiento 1987; Sarmiento et al. 1985). Biomass allocation is also variable between tree species, particularly during the first growth stages (Moreira 1992). Trees also differ in their responses to burning (Moreira 1992).

9.3 Responses of Savanna Species to Changes in Fire Regime

9.3.1 Grass Populations

Changes in the frequency of fires can affect grass populations on two accounts: (1) direct mortality, especially of young plants; (2) increased radiation reaching the soil surface, since fire removes the dry standing matter. Savanna grasses differ in their sensitivity to these two effects. The seedlings of some species are more shade-tolerant than others, and hence these species are capable of withstanding longer periods without a fire. Some species are more resistant to burning, and hence they persist under a regime of frequent fires. In the Venezuelan savannas, seedlings and small plants of *Andropogon semiberbis* are less tolerant to the shading generated by fire exclusion than those of *Sporobolus cubensis*. On the other hand, the seedlings and young plants of *S. cubensis* are more sensitive to the effects of fire as a mortality agent than those of *A. semiberbis* (Silva et al. 1990). Populations of these species will react differently to changes in fire frequency, since both rely on seed production and germination for population growth.

Annual and perennial grass species differ in their response to changes in fire frequency. The annual grass *Andropogon brevifolius* seems able to persist under annual fire frequency equal or higher than 0.29. This is equal to a fire every third year (Canales et al. 1994). In contrast, the perennial *A. semiberbis* cannot persist under fire regimes with annual frequency below 0.83, that is exclusion of fire every sixth year (Silva et al. 1991). The perennial *Sporobolus cubensis*, which, as discussed above, is more tolerant to shade and less tolerant to fire than *A. semiberbis*, probably tolerates fire frequencies lower than 0.83. Other common grass species may also differ in their responses to fire frequency, but there is still very little published information available.

Table 9.1.A Rate of increase in tree density (N) after fire exclusion in 1961 adjusted on a yearly basis for comparison. Calculated from San José and Fariñas (1991) as $N(t2) - N(t1)/ N(t1) [t2 - t1)]$; "tl" and "t2" are consecutive sampling years. **B** Rate of increase of relative density (P) of grass species in the open grassland. Calculated from San José and Fariñas (1991) as $P(t2) - P(t1)/ P(t1) \times [t2 - t1]$

	1969	1977	1983	1986
A				
In Open Grassland				
Curatella americana	1.14	0.15	0.78	0.14
Byrsonima crassifolia	0.45	0.11	0.14	0.004
Bowdichia virgilioides	-0.02	-0.01	0.61	0.004
Cochlospermum vitifolium		0.30	0.61	0.02
Godmania macrocarpa			16.17	0.04
Cordia hirta			3.29	-0.002
Machaerium pseudoacutifolium				0.25
In Groves				
Curatella americana	0.06	0.05	0.27	0.12
Byrsonima crassifolia	0.03	0.01	0.07	0.01
Bowdichia virgilioides	0.02	0.05	-0.01	0.01
Cochlospermum vitifolium	4.57	0.12	1.00	0.07
Godmania macrocarpa	1.86	0.09	1.21	0.09
Cordia hirta	9.00	0.27	0.27	-0.001
Machaerium pseudoacutifolium		0.25	0.94	0.22
B				
Trachypogon plumosus	-0.02	-0.06	-0.07	0.20
Axonopus canescens	0.08	0.09	-0.04	0.05
Hyparrhenia rufa		2.38	0.36	0.10

One additional source of information comes from long-term fire exclusion experiments. Data from the Biological Station in Calabozo, Venezuela (San José and Fariñas 1983, 1991) show different responses to fire exclusion from *Axonopus canescens*, *Trachypogon plumosus*, and *Hyparrhenia rufa* (Table 9.1). *Axonopus canescens* increased during the first 16 years of exclusion, whereas during the first 22 years of exclusion *T. plumosus* decreased steadily. *Hyparrhenia rufa* started to invade the area after 8 years, and increased rapidly thereafter. Changes in species presence may be due to the removal of direct mortality effects on seedlings, particularly important for seed relying species such as *A. canescens* and *H. rufa*. Also the decrease in light intensity reaching the ground seemed to have a stronger negative effect on *T. plumosus* than on the other species (San José and Fariñas 1983). Since we already know that *T. plumosus* is a strong competitor (Raventós and Silva 1995), it is probable that its decrease is favoring the increase of the other species. After prolonged fire exclusion, shifts in population growth occur, as shown in Table 9.1. Additional factors operating in this case may be rainfall, soil, and nutrient status.

The importance of the species composition for savanna stability is reinforced by a very transient seed bank in the soil, since seeds germinate readily and have limited survival in the field (Silva and Ataroff 1985; O'Connor and Pickett 1992). The seeds produced by precocious and early flowering grasses germinate as soon as they reach wet ground. The seeds produced by intermediate and late flowering species stay dormant on the surface of the soil throughout the year and will germinate at the beginning of the following rainy season. The seeds of tree species probably germinate during the first weeks of the wet season, although there are no detailed studies on the dynamics of tree seed banks in the field. Seeds that do not germinate do not survive and there is no apparent permanent seed bank in the savanna soil. Consequently, seedling recruitment depends on current seed production, and this in turn depends on the size of the population, the fraction of individuals blooming and their productive performance in that particular year. Further studies are needed to provide information on these aspects.

9.3.2 Tree Populations

Tree density increases in response to a decrease in fire frequency as has been shown in several exclusion experiments (Menaut 1977; San José and Fariñas 1983; Coutinho 1990). Furthermore, new tree species invade the savanna during fire exclusion, increasing the diversity of the tree layer. However, the various tree populations react differently to the change in fire regime, and therefore the savanna response to this perturbation depends on the initial composition of the tree layer. Propagule availability from neighboring forests is also important.

In the Calabozo's exclusion experiment mentioned above, San José and Fariñas (1991) recorded differences in population growth between and within sclerophyllous-evergreen and mesophyllous-deciduous tree species (Table 9.1). In the open savanna, evergreen trees increased rapidly in numbers after exclusion, whereas deciduous trees exhibited a long lag phase of at least 16 years. In the groves, in contrast, the deciduous trees increased strikingly during the first 8 years after exclusion, whereas the evergreen trees changed little. Among the sclerophyllous trees, *Curatella americana* showed a high growth rate compared to *Bowdichia virgilioides* (Table 9.1). Among the deciduous trees, *Cochlospermum vitifolium* needed only 8 years of exclusion to start increasing in the open grassland, whereas all the other deciduous species needed much longer.

These results are not easy to interpret based only on the effects of fire exclusion. Deciduous trees are little resistant to fire, especially in the early growth stages; therefore they are unable to invade the frequently burnt grassland, and are restricted to small groves protected from burning. After exclusion of fire, these populations are expected to rapidly colonize the grassland. However, fast-growing trees with wind-dispersed seeds such as *Cochlospermum vitifolium* and especially *Godmania macrocarpa*, increased very slowly in the fire-free grassland. In the 1983 census (22 years of exclusion), both species showed a burst of increase, especially *G. macrocarpa*, which during the 6 years between 1977 and 1983 increased 98 times in number (from 6 to 588 stems; San José and Fariñas 1991).

One plausible explanation for this lag is that in the period 1977-1983 other disturbances took place and acted synergically with the lack of fire to promote or restrict tree populations' growth. Alternatively, some changes in the nutrient status of the soil may be needed in the open grassland but not in the groves, for the trees to become established. In this regard, Fölster (1986) suggested that nutrient impoverishment by recurrent fires following deforestation and establishment of the grassland is the main cause deterring the recovery of the forest in the Gran Sabana (southern Venezuela). However, groves and open grasslands from the Orinoco Llanos do not seem to be significantly different in soil nutrient status (Sarmiento 1984). In any case, the Calabozo exclusion experiment shows that all tree species reacted positively to fire exclusion, but differ in the speed and magnitude of their reactions (San José & Fariñas 1991).

9.4 Responses to Changes in PAM Regime

The seasonal pattern in plant-available moisture (PAM) is determined by the rainfall regime. In turn, soil characteristics (topographic position, structure, and texture) influence water availability throughout the year; however, they change on a geomorphologic time scale, except when erosive processes are accelerated by land use. Consequently, short-term perturbations in PAM regime do not originate in the soil but rather result from changes in rainfall. These short-term changes seem to be common, although there is little published information on climate trends in savannas to document it. In southern Africa, pulses in rainfall are responsible for changes in the size of populations and the structure of savanna communities on different soils and are considered an overriding factor in several studies on the effects of grazing and fire (O'Connor 1985). A conclusion from the O'Connor analysis of southern Africa experiments is that drought spells induce important changes in savanna composition, and these effects increase with aridity. On the other hand, savanna resilience to drought (resilience defined as the ability of the system to return to the previous condition after a disturbance, Harrison 1979) seems to be high. Another important conclusion in O'Connor (1985) is the influence of tree density on the changes in the grass layer. When the woody/grass ratio (density of trees) was stable, the grassland component was not affected by overgrazing.

However, when changes in tree density took place during the experiment, the grass layer was very sensitive to grazing and overgrazing, especially during critical drought years. This implies a positive feedback for the whole system, since stability of one component (trees) results in added stability of another component (grasses). Unfortunately, these are very preliminary conclusions, and to my knowledge there are no published studies relating rainfall variations with savanna structural changes.

Several studies enforce the conclusion that savanna community is primarily responding to PAM regime (Medina and Silva 1990; Solbrig 1990; Teague and Smit 1992). Some results emphasize the differences in plant response to water stress in grasses and trees from seasonal savannas (see Chap. 6). Savanna grasses are tolerant to negative leaf water potentials. However, they differ in the degree of tolerance to water stress (Goldstein and Sarmiento 1987). Furthermore, grass species are distributed differentially along humidity gradients reflecting differences in their responses to the length of the dry season and to water logging (Silva and Sarmiento 1976a,b). These preferences are seemingly related to the phenological behavior of the population. Sarmiento (1983b) suggested that change in the length of the dry season results in increased representation of some phenological groups at the expense of other groups. Although Sarmiento's study was not conclusive, it provides support for this hypothesis. He concluded that early onset of rains would favor species blooming very early in the wet season ('precocious species'), whereas a prolonged rainy season would favor species blooming at the end of the wet season („late species"); a shortening of the rainy season was detrimental to both types but favorable to species blooming in the middle of the wet season („early" and „intermediate" species).

Trees are probably more sensitive than grasses to changes in water availability. Wet spells seem to be responsible for an increase in tree density in the dryer savannas of southern Africa (O'Connor 1985). Quaternary fluctuations in annual rainfall are related to major changes in tree densities in savannas and in the forest-savanna boundaries in South America (van der Hammen 1974, 1983) and Madagascar (Burney 1993). To what extent the response of the woody component to changes in water availability throughout the year depends on the species composition is not known. Evergreen trees with high underground biomass and large leaves (e.g., *Curatella americana*) depend on continuous water supply from the subsoil. Extended dry seasons for several consecutive years will have a negative effect on the growth of these trees due to progressive depletion of subsoil water. Goldstein and Sarmiento (1987) suggested that under these conditions, deciduous trees or trees with smaller leaf size and higher tolerance to drought should be favored.

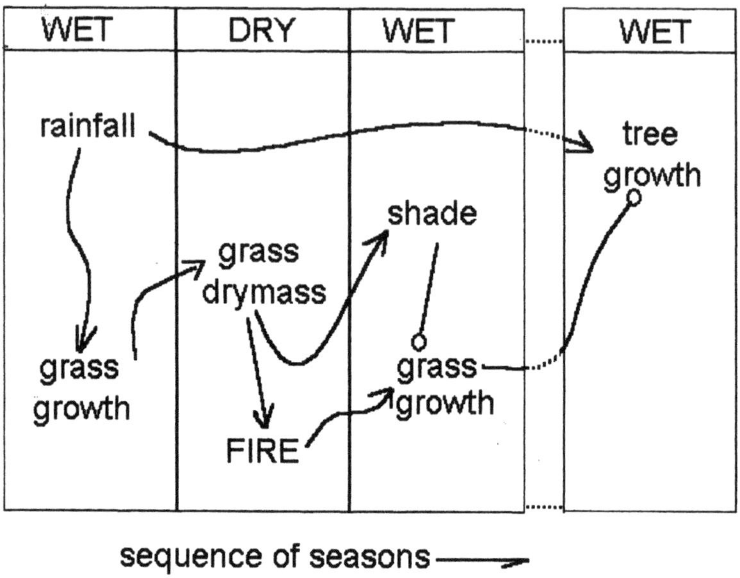

Fig. 9.1. Diagram showing possible interactions between savanna determinants (rainfall, fire) and savanna components (grass biomass and tree biomass) throughout a sequence of seasons. An *arrow* indicates positive effect; a *circle* means negative effect (see text for explanation)

Perturbations are not totally independent. On the contrary, there are strong interactions among rainfall, biomass, fire frequency and intensity, grazing, and nutrient fluxes in seasonal savannas. A pulse of increased annual rainfall will increase grass biomass during that season, and this will result in an increase in the amount of standing dead biomass during the next dry season. As a consequence, the probabilities of burning also increase. Alternatively, if there is no fire, shading will increase during the following wet season, reducing grass growth (Fig. 9.1). The consequences of a higher rainfall pulse may not be restricted to the short term. Tree growth and recruitment will increase, especially if fire does not occur. This will alter the tree/grass ratio in following years, affecting grass growth and reproduction and influencing fire and grazing regimes.

9.5 Conclusions

Savanna plants differ widely in their responses to changes in fire and humidity regime. The direction and extent of the changes the savanna community undergoes as a consequence of environmental perturbations depends on what species are facing the perturbation. These "initial conditions" in savanna community structure may be more important than species richness or other diversity index, and have to be considered to study the dynamics of savannas as nonlinear systems (Nicolis 1992). The importance of the species composition for savanna stability is reinforced by the transient seed bank in the soil. Although it has been claimed that functional stability of experimental grasslands perturbed by a drought spell depended on species richness per se (Tilman and Downing 1994), the response in this case was clearly tied to the presence of particular, drought-resistant species in the richer communities.

The diversity of savanna plant responses to environmental perturbations may explain the persistence of savanna communities within a broad range of environmental variation. Current knowledge suggests that savanna persists under the control of seasonal climate, fluctuating within certain bounds as the result of the occurrence of climatic disturbances and its concatenated effects on fire, grazing, and other factors.

Savanna studies have emphasized the similarities among species and classified them into functional groups, such as evergreen-sclerophyllous trees or perennial bunch grasses. To better understand savanna responses to perturbation, it is convenient to shift the focus to the comparative analysis of species morphology and their functional differences. The initial hypothesis of this chapter can be used to further research on the role these differences are playing in the stability of savanna communities.

Acknowledgments.
I thank Guillermo Sarmiento and Aura Azócar for suggestions on an earlier draft of this manuscript and Otto T. Solbrig for valuable criticism and suggestions. This research is being partially funded by CDCHT-ULA grant C-344-90. The chapter was written during my stay at Harvard University as a Charles Bullard Fellow.

References

Burney DA (1993) Late Holocene changes in Madagascar. Quat Res (Orlando) 40:98-106

Canales MJ, Silva JF, Caswell H, Trevisan MC (1994) A demographic study of an annual savanna grass (*Andropogon brevifolius* Schwarz) in burnt and in burnt savanna. Acta Oecol 15:261-273

Connell JH, Sousa WP (1983) On the evidence needed to judge ecological stability or persistence. Am Nat 121:789-824

Coutinho LM (1990) Fire in the ecology of the Brazilian cerrado. In: Goldammer JG (ed) Fire in the tropical biota. Springer, Berlin Heidelberg New York, pp 82-105

Ernst WHO, Kuiters AT, Tolsma DJ (1991) Dormancy of annual and perennial grasses from a savanna of southeastern Botswana. Acta Oecol 12: 727-739

Fölster H (1986) Forest-savanna dynamics and desertification process in the Gran Sabana. Interciencia 11(6):311-316

Goldstein G, Sarmiento G (1987) Water relations of trees and grasses and their consequences for the structure of savanna vegetation. In: Walker BH (ed) Determinants of tropical savannas. IRL Press, Oxford, pp 13-38

Harrison GW (1979) Stability under environmental stress: resistance, resilience, persistence, and variability. Am Nat 113:659-669

Holling CS (1986) The resilience of terrestrial ecosystems: local surprise and global change. In: Clark WC, Munn RE (eds) Sustainable development of the biosphere. Cambridge Univ Press, pp 292-317

Medina E, Silva JF (1990) The savannas of northern South America: a steady state regulated by water-fire interactions on a background of low nutrient availability. J Biogeogr 17:403-413

Menaut JC (1977) Evolution of plots protected from fire for 13 years in a Guinea Savanna of Ivory Coast. Actas IV Simp Int Ecol Trop. Impresora de la Nación, Panamá, pp 540-558

Moreira AG (1992) Fire protection and vegetation dynamics in the Brazilian Cerrado. PhD Dissertation, Harvard Univ

Mott JJ, Andrew MH (1985) The effect of fire on the population dynamics of native grasses in tropical savannas of north-west Australia. Proc Ecol Soc Aust 13:231-239

Nicolis G (1992) Dynamical systems, biological complexity and global change. In: Solbrig OT, van Emdem HM, van Oordt PGWJ, (eds) Biodiversity and global change. IUBS Paris, pp 21-32

O'Connor TGO (1985) A synthesis of field experiments concerning the grass layer in the savanna regions of southern Africa. SANSP # 114. Found Res Dev, Pretoria

O'Connor TGO (1993) The influence of rainfall and grazing on the demography of some African savanna grasses: a matrix modeling approach. J Appl Ecol 30:119-132

O'Connor TGO, Pickett GA (1992) The influence of grazing on seed production and seed banks of some African savanna grasslands. J Appl Ecol 29:247-260

Oliveira PE, Moreira AG (1992) Anemocoria em espécies de cerrado e mata de galeria de Brasilia, DF. Rev Bras Bot 15:163-174

Oliveira PE, Silva JCS (1993) Reproductive biology of two species of *Kielmeyera* (Guttiferae) in the cerrados of central Brazil. J Trop Ecol 9:67-79

Raventós J, Silva JF (1988): Architecture, seasonal growth and interference in three grass species with different flowering phenologies in a tropical savanna. Vegetatio 75:115-123

Raventós J, Silva JF (1995) Competition effects and responses to variable number of neighbors in two tropical savanna grasses in Venezuela. J Trop Ecol 11:39-52

San José JJ, Fariñas MR (1983) Changes in tree density and species composition in a protected *Trachypogon* savanna, Venezuela. Ecology 64:447-453

San José JJ, Fariñas MR (1991) Temporal changes in the structure of a *Trachypogon* savanna protected for 25 years. Acta Oecol 12:237-247

Sarmiento G (1983a) The savannas of tropical America. In: Bourlière F (ed) Tropical savannas. Elsevier, Amsterdam, pp 245-288
Sarmiento G (1983b) Patterns of specific and phenological diversity in the grass community of the Venezuelan tropical savannas. J Biogeogr 10:373-391
Sarmiento G (1984) The ecology of neotropical savannas. Harvard Univ Press, Cambridge, 235 pp
Sarmiento G, Monasterio M (1983) Life forms and phenology. In: Bourlière F (ed) The tropical savannas. Elsevier, Amsterdam, pp 79-108
Sarmiento G, Goldstein G, Meinzer F (1985) Adaptive strategies of woody species in neotropical savannas. Biol Rev 60:315-355
Silva JF (1987) Responses of savannas to stress and disturbance: species dynamics. In: Walker BH (ed) Determinants of tropical savannas. IRL Press, Oxford, pp 141-156
Silva JF, Ataroff M (1985): Phenology, seed crop and germination of coexisting grass species from a tropical savanna in western Venezuela. Acta Oecol Oecol Plant 6:41-51
Silva JF, Sarmiento G (1976a) La composición de las sabanas de Barinas en relación con las unidades edáficas. Acta Cient Venez 27:68-78
Silva JF, Sarmiento G (1976b) Influencia de factores edáficos en la diferenciación de las sabanas. Análisis de componentes principales y su interpretación. Acta Cient Venez 27:141-147
Silva JF, Raventós J, Caswell H (1990) Fire, fire exclusion and seasonal effects on the growth and survival of two savanna grasses. Acta Oecol 11:783-800
Silva JF, Raventós J, Caswell H, Trevisan MC (1991) Population responses to fire in a tropical savanna grass: a matrix model approach. J Ecol 79:345-356
Solbrig OT (ed) (1990) Savanna modeling for global change. Biol Int Spec Issue 24, 47 pp
Teague WR, Smit GN (1992) Relations between woody and herbaceous components and the effects of bush clearing in southern African savannas. J Grassl Soc South Afr 9:60-71
Tilman D, Downing JA (1994) Biodiversity and stability in grasslands. Nature 367:363-364
Van der Hammen T (1974) The Pleistocene changes of vegetation and climate in tropical South America. J Biogeogr 1:3-26
Van der Hammen T (1983) The palaecology and palaeogeography of savannas. In: Bourlière F (ed) Tropical Savannas. Elsevier, Amsterdam
Veenendaal EM, Shushu DD, Scurlock JMO (1993) Responses to shading of seedlings of savanna grasses (with different C_4 photosynthetic pathways) in Botswana. J Trop Ecol 9:213-229
Walker BH, Noy-Meir I (1982) Aspects of the stability and resilience of savanna ecosystems. In: Huntley BJ, Walker BH (eds) Ecology of tropical savannas. Springer, Berlin Heidelberg New York
Walker BH, Ludwig D, Hooling CS, Peterman RM (1981) Stability of semiarid savanna grazing systems. J Ecol 69:473-498

Summary and Areas for Future Research

10 Biodiversity As Regulator of Energy Flow, Water Use and Nutrient Cycling in Savannas

10 Biodiversity As Regulator of Energy Flow, Water Use and Nutrient Cycling in Savannas

Zdravko Baruch, A. Joy Belsky, Luis Bulla, C. Augusto Franco, Irene Garay, Mundayatan Haridasan, Patrick Lavelle, Ernesto Medina, and Guillermo Sarmiento

10.1 Introduction

A central question of the Diversitas program, Brasilia (1993), was how the reduction of biodiversity, or more precisely species richness in a given ecosystem, will affect the processes characterizing its functioning in those aspects related to energy and matter flow, to reproduction and perpetuation in time, and to resistance and resilience in the face of disturbances of variable intensities.

In order to discuss the role of biodiversity on biogeochemical cycles in savanna ecosystems, it is necessary to define the systems we are dealing with. This definition includes aspects of "savanna structure" and "savanna function" and is broad enough to include the ecosystems that are heuristically referred to as savannas, while adding constraints that provide boundaries on our definition.

10.2 Savanna Structure

A savanna is a structurally simple but spatially patchy tropical ecosystem characterized by a herbaceous layer dominated by xeromorphic C_4 grasses and, in most cases, having a woody component consisting of deciduous or evergreen trees or shrubs. Savanna composition and

Ecological Studies, Vol. 121
Solbrig, Medina and Silva (eds) Biodiversity and Savanna Ecosystem Processes

structure vary both spatially and temporally as the height and density of the woody components change in response to fire, herbivory, nutrient availability, or climate.

10.3 Savanna Function

Savannas are ecosystems characterized by relatively low biomass compared to forests. This low biomass may result from a variety of naturally occurring factors: low amounts of plant-available moisture (PAM), low concentrations of plant-available nutrients (PAN), shallow soil depth, recurrent fire, or intensive herbivory. Plant biomass is further restricted by the strong seasonality of tropical climates, which reduces the activity and/or leaf area of many species during part of each year, and by the dramatic effects of fire and herbivory on above-ground biomass and below-ground mineralization processes. Of unique importance to savanna ecosystems are their below-ground systems, which serve as energy and nutrient reservoirs that protect individual plants and entire ecosystems from recurrent perturbations such as drought, fire, and herbivory.

10.4 Biogeochemical Cycles in Savannas

Biogeochemical cycles have been extensively documented for different ecosystems; however, information on savannas is far from exhaustive (Frost et al. 1986; Walker 1987). The mechanisms involved in the processes of organic matter production, water and nutrient cycling, and decomposition are well understood, even though their quantitative aspects have not been worked out satisfactorily (Menaut et al. 1985; Goldstein and Sarmiento 1987; Medina and Silva 1990; Medina and Bilbao 1991). A scheme of the complexity of processes that are defined as biogeochemical cycling in savanna ecosystems must incorporate primary production, water uptake and transpiration, nutrient uptake, and organic matter decomposition as primary variables, as well as biomass-allocation patterns, herbivory, and interactions among all these processes (Table 10.1, following in part the descriptions of Main 1992 and Hobbie et al. 1993).

Table 10.1 Schematic description of the processes involved in biogeochemical cycles in savanna ecosystems.

Biogeochemical cycle	Processes involved
Energy and carbon fixation	Photosynthesis Allocation of biomass for leaf area development
Water cycling	Water uptake and transpiration by primary producers Allocation for: leaf area development, root biomass and area
Nutrient cycling	Nutrient uptake by primary producers Roots Symbiosis and mutualisms Rhizosphere Mycorrhiza *Rhizobium* symbiosis *Frankia* symbiosis
	Nutrient transfer and redistribution Living plant matter consumption (herbivores) Dead plant matter consumption (detritivores) Secondary consumers
	Nutrient release Decomposition processes (soil microorganisms) Mineralization
	Soil formation Organic matter conditioning and humification
Interactions	Organic matter production requires nutrient and water uptake, while water cycle in the system introduces nutrients into, and leaches nutrients out of, the system

Savanna ecosystems are characterized by a number of structural and functional features that may have significant bearing on the efficiency and stability of biogeochemical cycles:

1. The coexistence of trees and grasses in a dynamic equilibrium that is regulated by water availability and fire regimes has strong implications for the dynamics of the system regarding light interception, water balance, and layering of soil-resource utilization (Walker and Noy-Meir 1982).
2. Within the herbaceous matrix, the cooccurrence of grasses and sedges, having in general a C_4 photosynthetic pathway, and forbs, having in general a C_3 photosynthetic pathway, results in patchiness of herbaceous

layer productivity and water- and nutrient-use efficiency (Medina 1982; Sarmiento 1984). In addition, the diversity of phenological types (early and late growers, annuals and perennials) provides a temporal dimension, which allows primary productivity to take place throughout the year (Sarmiento 1983).

3. The occurrence of nitrogen-fixing organisms, both free-living microorganisms (Cyanophyceae and bacteria) and symbiotic higher plants (rhizobial symbionts), creates further spatial heterogeneity in nutrient distribution, particularly that of nitrogen and calcium (Medina and Bilbao 1991). In addition, the role of widespread mycorrhizal symbiosis for water balance and phosphorus uptake in savanna plants has not been properly documented yet.
4. Root/shoot ratios in savanna ecosystems, particularly within the herbaceous layer, are considerably greater than 1, a feature providing resistance to stress and disturbance from drought, fire, and herbivory (Sarmiento 1984; MacNaughton 1985; Frost et al. 1986).
5. Interactions of fire, herbivory by large animals, and the activity of ants and termites in nutrient conservation and cycling constitute a unique feature of savannas that requires precise documentation and modeling.

According to Table 10.1, there are a number of points in which changes in biodiversity could modify both quantitatively and qualitatively the pattern of biogeochemical cycle in a given ecosystem. Biogeochemistry is essentially determined by the development of biological surfaces, i.e., photosynthetic surfaces chemically fixing incoming sunlight and absorbing CO_2 (primary production of organic matter) and water- and nutrient absorbing surfaces. Primary production is proportional to the amount of light energy intercepted, which is a function of shoot development and architecture, and the intrinsic capacity of the biochemistry of the photosynthetic apparatus to incorporate CO_2 into organic compounds, which is greatly dependent on nutrient availability. Moreover, because of the stomatal control of both water loss (transpiration) and CO_2 uptake, these processes are intimately related. The ratio of water consumed in transpiration per unit of organic matter produced is a valuable index of water-use efficiency at the ecosystem level. Development and spatial distribution of photosynthetic surfaces depend on the morphology of the species. In savannas, predominance of grasses and forbs is associated with dense packing of photosynthetic surfaces that are located near the soil surface, while predominance of trees results in a vertical distribution of light-intercepting surfaces leading to a more efficient process of energy capture. Efficiency of water- and nutrient use for primary production is also species-dependent.

Nutrient and water absorption depend on development of root biomass and the effective surface of interaction with the soil matrix. Patterns of biomass allocation, root depth, and efficiency of symbiotic associations also vary with the species composition of the primary producers. Primary producers with different habits tend to differ substantially in the quality of organic matter they produce (i.e., carbon:nitrogen and lignin:nitrogen ratios; proportion of protein, lipids, and carbohydrates). Therefore, changes in species composition in a given ecosystem may result in changes of patterns of herbivory and in nutrient sequestering by the organic-matter decomposers within the soil matrix.

There are several examples in the literature documenting the potential and actual effects of certain species on the rates of processes determining biogeochemical cycling in natural and disturbed ecosystems. Some species might be particularly efficient in recycling nutrients that are critical for ecosystem functioning (Muller and Bormann 1976), while the introduction of some species can open different pathways for ecosystem succession. That has been the case with the introduction of nitrogen-fixing species (Vitousek et al. 1987) or the invasion of fire-resistant grasses in Hawaii (Hughes et al. 1991). The former improved nutritional conditions in the soil, allowing the survival of more nitrogen-demanding species, while the latter increased the fire risk in non-fire-resistant systems.

10.5 Functional Groups

Species occurring in any given ecosystem are differentiated according to their morphological and physiological characteristics. Differentiation between species may be substantial among species growing in high-stress (resource poor or severely and/or frequently disturbed) ecosystems and subtle in resource-rich, low-stress ecosystems. In semiarid ecosystems the rate of species differentiation and extinction is higher than in mesic ecosystems (Stebbins 1952). Therefore, at a given time, semiarid systems sustain lower species diversity than mesic ones, although that is not necessarily true in a historic perspective. In highly stressed ecosystems, resource availability limits the number of cooccurring species with similar ecological requirements. Only those species highly adapted to the stressing factor survive. Higher availability of resources in low-stressed ecosystems allows the packaging of more species with similar ecological requirements within a certain space and time. These considerations have to take into account species richness and diversity, which are a function of the area, age, and evolutionary history of the habitat, as well as the size and environmental requirements of cooccurring species. Another aspect necessary for understanding the relationship between stress and biodiversity is that the nature of stress is multiple in ecological settings (Chapin 1991). For example, drought stress frequently leads to disturbances in nutrient acquisition, and occurs under conditions of high irradiation and possibly high temperature.

Different combinations of stresses may lead to widely different responses at the ecosystem level, resulting in variable numbers of species.

In principle, species can be ordered according to their role and relative importance in ecosystem processes. Groups of species that are classified on the basis of their morphophysiological and phenological properties and have possibly "similar" impacts on ecosystem processes are called functional groups (Hobbie et al. 1993; Vitousek and Hooper 1993; for a more extensive and recent discussion see Huston 1994). There is a standing controversy on the similarity of functions of species classified within a certain functional group. For instance, primary producers in a savanna represent a complex functional group consisting of species of different habit, size, and physiological requirements. The subdivision of the primary producers into morphological types such as herbs and trees also results in complex groups, when physiological traits and other morphological characteristics are taken into account. The herbs can be further separated into grasses and forbs, C3 and C4 photosynthetic types, nitrogen-fixing and nonfixing, deep- and shallow-rooted, early and late growers, and so on. However, in many savanna types the number of species that can be attached to a certain functional group (i.e., primary producers, consumers, decomposers, etc.) is large, therefore the question of species equivalence (or redundancy) within a given functional group has scientific significance (Walker 1992). The probability of finding equivalent species within functional groups that are important for biogeochemical cycling in a savanna ecosystem is high, but it does not mean that those species are also redundant regarding other aspects of ecosystem function, for example in regard to stability and resilience (Lawton and Brown 1993).

Definition of functional groups depends on which process is being analyzed. Clearly, a given species may belong to several functional groups; and to a certain extent, species belonging to more than one functional group may be more critical for ecosystem function than are species restricted to a single group. In addition, functional groups with a large number of species are characterized by a larger number of species interactions, which might be of significance in strongly fluctuating environments or in environments having a high frequency of disturbances. Both characteristics apply to savanna ecosystems and should be remembered when analyzing the importance of a certain species in biogeochemical cycling (Frost et al. 1986).

Here only a few examples of functional groups in savannas will be given, including those of primary producers, megafaunal herbivores, and soil invertebrates. A high priority should be given to defining functional groups within the decomposer community, because very little is known about their population properties and physiological requirements. Prediction of the response of savanna ecosystems to changes in macroorganism diversity will depend on the understanding of the impact of physicochemical stress and changes in substrate quality on the proportion, abundance, and activity of soil microorganisms.

10.6 Primary Producers

Plant functional groups can be defined according to habit and size. In savanna ecosystems, it is important to distinguish between herbaceous and ligneous plants, because these plants are related to different rates of organic matter production and accumulation and to patterns of organic matter allocation to above- and below-ground organs. It has been shown that there is a strong correlation between degree of woodiness (and plant size) and the availability of resources (water and nutrients). As the availability of nutrients and water increases, the number of woody species, particularly trees, increases. Under conditions of light limitation, trees are more competitive because of the vertical displacement of their photosynthetic area. In addition to this general tendency, tree/grass ratios are influenced by intensity of herbivory and fire regimes (Belsky 1990; Medina and Silva 1990). In both cases, above-ground biomass is particularly affected. However, while fire impact is generally restricted to the dry season, and is neutral in destroying above-ground dead grass biomass and canopies of evergreen trees, herbivory occurs primarily during the rainy season, is selective, and is accompanied by the effect of trampling. The effect of herbivores is not necessarily unidirectional; they may either increase or reduce the tree/grass ratio of a given savanna, depending on the degree of environmental stress and their selectivity. Another important difference between fire and herbivory is that fire causes a relative homogeneous volatilization of organic matter and certain nutrients (N, S, K), while herbivory leads to nutrient relocalization and patchiness. The comparatively large root/shoot ratios in savannas minimize the loss of nutrients due to burning during the dry season.

A scheme of the distribution of functional groups of primary consumers has been devised following suggestions of Hobbie et al. (1993; Fig. 10.1). The scheme hypothesizes that the dominant type of functional group among primary producers will be related to the availability of resources (water and nutrients) (Tilman 1990) and will be modulated by the impact of fire and herbivory (Medina and Silva 1990). As the availability of resources increases, numbers of trees increase, resulting in greater competition for light (Schulze and Chapin 1987; Hobbie et al. 1993). As stated before, establishment of the equivalence of the species within each functional group is of paramount importance in order to determine if they can be replaced under certain circumstances. Substitutability may be measured as the compensation by density increase of some species, after elimination of one or more species within a given functional group (Walker 1992). The establishment of interspecific equivalence requires both spatial and temporal dimensions, therefore the analysis should take into account morphology, nutritional requirements, intrinsic relative growth rate (RGR), productivity, and phenology.

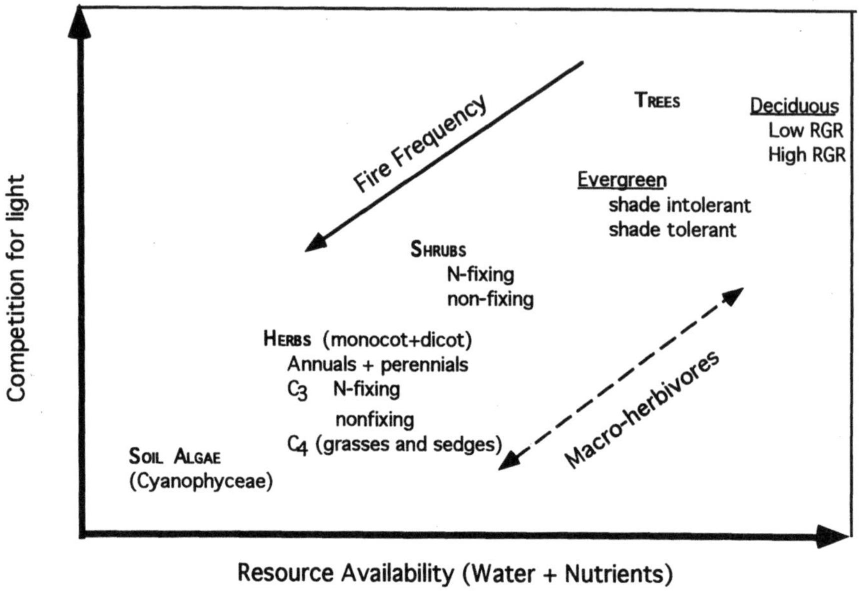

Fig. 10.1. Variation in the predominance of functional groups of primary producers (≈ lifeforms arranged according to size and ecophysiological properties) (in analogy to Hobbie et al. 1993)

10.7 Faunal Functional Groups

Examples of functional groups of savanna macroherbivores are defined according to their diet:

Grazers: feed mostly on the herbaceous layer.
Browsers: feed mostly on the shrub and tree layers.
Granivores: feed mostly on plant seeds.
Nectivores: feed mostly on nectar produced in floral and extrafloral nectaries.
Frugivores: feed mostly on fruits.

10.8 Soil Fauna

The soil fauna plays a critical role in processing plant residues, which are subsequently incorporated into soil organic matter (SOM) or rendered more accessible to decomposition (mineralization) through soil microorganisms. Lavelle (1987) distinguishes three major groups of soil macroinvertebrates:

1. Epigeics live in and feed on litter. As a result of their feeding activities, they produce fecal pellets in which accelerated release of nutrients may be observed at short scales of time; in the longer term, mineralization of soil organic matter may be slowed down because the initial flush of mineralization rapidly comes to an end as a result of reduced oxygen supply and porosity.
2. Anecics export organic matter from the litter system to other decomposition systems, i.e., to the subsoil itself (anecic earthworms) or to nests of social insects. There, organic matter is eventually digested and the residues are mixed with mineral elements taken from different soil horizons. Structures created by these organisms include galleries, mounds, macropores and aggregates (micro- and macroaggregates).
3. Endogeics include different subgroups living in the soil and feeding on soil organic matter. Their activities result in the formation of casts, which are stable macroaggregates at the scale of years, and macropores. These structures affect SOM and water dynamics at different scales of time and space. In earthworm casts, for example, mineralization of SOM is strongly enhanced at the scale of hours to days, but at the scale of months to years SOM, is protected. After the initial flush of mineralization following deposition of the fresh cast, the compact structure of the cast soon decreases mineralization rates (Lavelle and Martin 1993).

Examples of the types of organisms classified within each group are listed in Table 10.2.

Table 10.2. Examples of functional groups of soil macroinvertebrates according to Lavelle (1987)

Invertebrate Group	Examples
Epigeics	Xylophages (termites and nonsocial arthropods) Leaf litter feeders (nonsocial arthropods, epigeic earthworms, epigeic termites)
Anecics	Burrowing earthworms Fungus growing + foraging termites Leaf cutting ants Diplopoda
Polyhumics endogeics	Humivorous termites Polyhumic earthworms
Mesohumic endogeics	Mesohumic earthworms
Oligohumic endogeics	Oligohumic earthworms

10.9 Relationship Between Biodiversity and Biogeochemical Cycling

10.9.1 Hypothesis Formulation

Developing hypotheses and experimental tests to measure the impact of biodiversity on the function of savanna ecosystems is inherently difficult. First, it demands an alteration in the way in which biodiversity is normally viewed. Whereas ecologists have traditionally asked how different ecosystem variables affect biodiversity, in this chapter we ask whether biodiversity has an impact on ecosystem function. In other words, we attempt to develop a hypothesis on how ecosystem processes, such as energy, water, or nutrient fluxes, respond to increases or decreases in species richness.

The goal of our discussion is to develop a set of nontrivial hypotheses that can be tested to determine whether biodiversity has an effect on the fluxes of energy, carbon, water, or nutrients through savanna ecosystems. In our considerations several basic guidelines are taken into account, although it is not always possible to follow them consistently:

- Biodiversity is a measure both of species richness (total number of species) and of structural diversity (i.e., leaf-area distribution, canopy structure, soil heterogeneity, etc.). For simplicity, our discussions deal mainly with the α-diversity of limited areas rather than the β-diversity over larger areas, although it is recognized that biodiversity needs to be examined at different spatial scales.
- Hypotheses on productivity should be divided into primary productivity and secondary productivity. These two levels may have differing relationships to biodiversity.
- Species richness and species equitability (evenness) should be kept separate in all discussions; ideally, hypotheses should be developed as parallel sets for both components of biodiversity.

We discuss three approaches for testing hypotheses: comparative, experimental, and use of natural experiments. Each approach offers a range of advantages and disadvantages.

Comparative Approach.
Biodiversity and ecosystem variables such as PAM, PAN, productivity, horizontal structure, fire, maximum and minimum temperatures, rainfall could be measured in a large number of savannas and the relative importance of biodiversity on productivity, nutrient use, water-use efficiency, etc. could be determined by multiple regression, factor analysis, or path analysis. The advantage of this approach is that it would facilitate

the study and comparison of savannas around the world. Therefore, it may allow the collection of data on little-known processes while avoiding the problems associated with disturbing plots (see below). The disadvantage of this technique is that the comparative approach requires the collection of large amounts of data, without the results being strongly predictive.

Experimental Approach.

Biodiversity could be manipulated by removing or adding individual species or functional groups of species from savanna plots and then measuring ecosystem responses such as changes in nutrient or water movement through the soil, rainfall interception, quantity of runoff, rainfall-use efficiency, etc. This approach has the advantage of holding most important variables constant while testing one factor at a time. It also allows straightforward modeling of ecosystem responses and strong inferences can be made while testing meaningful hypotheses.

The manipulation of biodiversity in natural savannas involves the development of whole-ecosystem experiments, and therefore the planning has to be based on long-term observations and measurements. Disturbances resulting from species removal, additions, or changes in environmental stresses (i.e., fire regime, water or nutrient availability, composition of the herbivore community), possibly leading to changes in species composition, dominance, and density, have to be documented on both a short- and a long-term basis. Species deletion can be performed either directly (disturbing the community in the short term) or indirectly through changes in the ecosystem regulating factors associated with the occurrence of a given species. The general type of experiment that could be conducted may be related to individual species or to groups of species that presumably play significant roles in the ecosystem. This knowledge would be obtained initially from a detailed analysis of ecosystem species composition and abundance, and could be tested using the following examples:

1. Species deletion using specific biocides for:
   - dominant primary producer,
   - most abundant herbivore,
   - soil invertebrates,
   - soil microflora and microfauna.
2. Species addition through:
   - dispersal of propagules of exotic species of various habits,
   - introduction of herbivore species,
   - inoculation of new soil microorganisms (for example, *Rhizobium* varieties or spores of mycorrhizal fungi).
3. Manipulation of environmental constraints:
   - modifications of water availability,
   - increases in nutrient availability,
   - modification of the fire regime.

Some disadvantages of this approach are:

1. The removal of rare species would most probably produce changes below the level detectable in natural systems having a normal degree of temporal and spatial variability. Therefore, studies consisting of the removal of rare species can be considered futile.
2. The removal of common or dominant species would definitely produce changes in ecosystem function, but changes related to the loss of these species may be confounded with the disturbance required to remove the species. For instance, the direct elimination of a certain species of primary producers involves either the removal of roots, which would disturb the soil, or the roots would be left in place after killing the organisms, enriching the soil column as they decompose. In addition, the time necessary for the system to equilibrate following species removal would be unknown. It is also important for investigators to distinguish between the effects of the disturbance and the effects of a reduction in biodiversity. In addition, it is not clear whether the removal of one common species would provide a measure of a change in biodiversity or simply a measure of the importance of that one species to the community. For example, would changes resulting from the removal of elephants from African savannas tell us about the importance of consumer biodiversity or simply about the function of elephants in savannas?
3. Removal of whole functional groups would, without doubt, alter ecosystem function; but once again, these results might inform us more about the importance of each functional group than the importance of high species diversity.

Natural Experiments.

Analyses of natural experiments (for example, the relatively recent invasion of African grasses into South American savannas) offer the advantage of a long period of equilibration so that communities will have adjusted to the perturbation. In addition, these natural experiments often occupy large and diverse areas, which would provide adequate replication and facilitate the measurement of response. Although this type of experiment has few pitfalls, the answers to several questions are already known; the invasion of African grasses, for example, has lowered biodiversity in South American savannas while increasing productivity. However, the impact of exotic grasses on the rate of biogeochemical cycling, or on the mechanisms that prevent reinvasion of native grasses remain as fruitful lines of inquiry in understanding the relationship between biodiversity and ecosystem function (Baruch and Fernández 1993).

10.9.2 Hypothesis Testing

Testing of any hypothesis regarding the regulatory effect of biodiversity on biogeochemical cycling will have to measure the fluxes of energy and matter through natural or disturbed ecosystems. For savanna ecosystems the methodology is comparatively well known and it involves the measurement of all relevant environmental parameters (light, rainfall, temperature, soil percolation, nutrient concentrations) as well as a description of ecosystem compartments (producers, consumers, decomposers, soil).

Standard measurements should provide a quantitative assessment of:

- Energy balance of the ecosystem (radiation input, output, and retention).
- Water balance: rainfall, evapotranspiration, run-off and percolation.
- Nutrient concentration in water fluxes and estimation of inputs, losses, and recirculation.
- Compartmentalization of biomass and nutrients and seasonal variations in primary and secondary production.
- Soil organic matter quality.
- Microbe community composition.
- Processes of nutrient transfer and release due to:
  - herbivory,
  - decomposition (comminution and mineralization rates, soil respiration).

Techniques for measurement of relevant ecosystem processes such as primary productivity, energy interception, and nutrient cycling have been adequately documented and described (see Bormann and Likens 1979; Walker and Menaut 1988; Pearcy et al. 1989; Hall et al. 1993). Methodology for the measurement of soil processes has been described in detail in the *Manual of Methods of the Tropical Soil Biology and Fertility Group* (Anderson and Ingram 1993). Therefore the hypotheses formulated include only a proposition considered to be true and a brief rationale to explain the extent and implications of the proposition. Specific predictions are indicated in those cases where testing procedures are not obvious from the rationale.

10.10 Specific Hypotheses on Species Diversity and Ecosystem Function

10.10.1 General

Hypothesis 1 Savanna ecosystems having high biodiversity will be better able to acquire and sequester limiting environmental resources than those of lower biodiversity.

Rationale Diverse ecosystems should be more efficient at acquiring and retaining limiting resources (rather than nonlimiting resources) because most of the species will be adapted for taking up and sequestering those resources. Reduction in species richness might cause the loss from the system of a fraction of those resources, since fewer species would be available to take up and hold the resources during all parts of the year. Ecologically similar (or redundant) species should protect the ecosystem from a loss of resources in case of species extinction, disturbance, normal environmental fluctuations, or extreme climatic conditions.

Hypothesis 2 The loss of species from ecosystems will affect the availability of resource for the remaining species (even if the resources are not lost), which may further alter species composition in the community.

Rationale The replacement of native species by African species in South America resulted in a reduction in biodiversity in these communities and may have altered the availability of nutrients, water, or energy, and initiated a new and progressive loss of species from the community. Another example is in the Great Basin of the United States, where *Bromus tectorum* has replaced many native herbaceous species, primarily by reducing the amount of water available to these species. As a result, the frequency of fires has increased, leading to further loss of native species. Species turnover and fire are most likely accompanied by losses in soil fauna and microorganisms, preventing recolonization by native species. Therefore, the high dominance of one introduced species, *B. tectorum*, has altered the ability of the ecosystem to recover its original species.

Hypothesis 3 Loss of biodiversity will have greater consequences at the landscape level than at the community level.

Rationale Species that may be rare or redundant in one habitat may be critical in another. Therefore, loss of certain species may have no effect on their local community but may result in a loss of resources in neighboring communities. For example, the grass *Cynodon plectostachys* occurs at low density and is probably redundant in open grassland-savannas in southern Kenya; however, it forms a monoculture under the crown of most tree species in the drier parts of the ecosystem. Loss of this species would probably result in the loss of nitrogen from subcrown habitats, which might affect tree growth.

Hypothesis 4 The removal of dominant species will have greater impact on ecosystem function in less diverse than in more diverse communities.

Rationale In highly diverse communities, the removal of a single dominant species should not result in a significant loss of important resources because there is some probability that other species in the community will increase in size or frequency and thereby capture the resources that have been released. In less diverse communities, there will be fewer similar species and perhaps none that are able to control the resources the same way.

10.10.2 The Role of Functional Groups in Maintaining Ecosystem Function

Hypothesis 5 Functional diversity in a savanna ecosystem minimizes loss of resources (energy, water and nutrients). Any change in functional diversity will decrease the amount of resources used, leaving some unutilized resources that eventually could be lost from the system.

Rationale If a functional group is totally eliminated from the system, some resources will not be captured and then their flux within the system will decrease. Thus, if trees are eliminated from a savanna, (a) total leaf area will decrease, resulting in less energy entering the system, (b) total root length will decrease, reducing both the amount of water transpired and the amount of mineral elements absorbed.

10.10.3 Relations Between Biodiversity and the Structure of the Primary Producers Functional Group

Hypothesis 6 Changes in biodiversity of primary producers that result in variations of system structure (biomass allocation, leaf area amount and distribution, etc.) affect water, nutrient, and energy flow.

Rationale Rates of water, nutrient, and energy cycling through ecosystems depend on the horizontal and vertical structural features of their primary producers, as represented by leaf area development, extension and area of the root system, and vertical stratification of the above-ground biomass.

These changes in structure most likely result from variations in the proportions of functional groups within primary producers. Examples of such changes are the modifications in the proportion of species with extensive *versus* those with intensive root systems (trees and grasses respectively), or of species with symbiotic associations (rhizobial symbionts and mycorrhizal associations). These structural changes may affect ecosystem function more than changes in species richness alone. The concept of species substitutability is builtin, primary producers with similar structure and physiology can substitute for one another.

Corollary 1.
Provided that the structure of the system as well as the proportion of the different structural elements are maintained, species can be removed without affecting the water, nutrient, and energy fluxes.

Corollary 2.
Patches of savannas that are vertically complex affect neighboring resources by creating suitable habitats for species with a wide range of resource requirements (light, water, and nutrients). Therefore, changes in vertical complexity will affect energy and material fluxes through the system.

Test 1. In a situation of equal proportions of trees and grasses, removal of the same proportion of leaf area in both groups will have a lesser effect on ecosystem function than the removal of the same leaf area from the grasses or the trees alone.

Test 2. Sites with different numbers of species in the tree and grass layers, but similar standing biomass of each group, should have similar fluxes of energy, nutrients, and water.

Test 3. Similar proportional changes in structural properties in systems with different proportions of tree-grass covers, should give a similar magnitude of changes in the fluxes, irrespective of species composition.

Hypothesis 7 Introduction of alien species of a certain functional group (grasses, trees, shrubs) will lead to changes in biogeochemical cycling according to the physiological characteristic of the species.

Rationale
a) Introduction of highly productive grasses affects productivity, temporal distribution of biomass production and reproduction (phenology), flammability of the above-ground biomass, and quality of biomass for herbivores

b) Introduction of nitrogen-fixing organisms increases patchiness of nutrient availability in the savanna, increases pasture quality for herbivores, and accelerates decomposition chains.
c) Evapotranspiration and water interception are higher while water percolation and runoff are slower in savannas dominated by native, slow-growing grasses compared to introduced fast-growing grasses, because of less cover and lower steady-state leaf conductance of native grasses.

10.10.4 Diversity of Underground Communities and Biogeochemical Cycles

Hypothesis 8 The diversity of soil macroinvertebrates determines plant production through the creation of structures that improve the efficiency of water and nutrient use at different scales of time and space.

Rationale Soil macroinvertebrates have developed different strategies to move and feed in the soil system. Each of these strategies results in the formation of complex structures in which the patterns and rates of nutrient cycling may be highly diverse (Lavelle 1987).

Test:

Comparative Approach.

Soil macrofauna communities show significant variations along rainfall gradients as well as biogeographical patterns. The experiment will consist of comparing the soil physical structure in soils with similar textures, but colonized by different groups of soil macroinvertebrates:

1. Surface features ("états de surface" following methodology of Valentin and Casenave 1992) and physical parameters like, e.g., water infiltration rate;
2. Soil aggregation as measured by Blanchart et al. (1990), if soil texture allows it;
3. Bulk density and the structure of porosity, including description of burrow systems and nests (e.g., Braudeau 1988).

Experimental Approach.

Experiments will include (1) the selective removal of functional groups that would suffer no other disturbance, and (2) the progressive addition of species in an artificial system (mesocosm in which plants would be cultivated in a previously sieved soil in which different soil fauna groups could be introduced; see, e.g., Blanchart et al. 1990; Spain et al. 1992).

In these experiments relevant parameters of the soil structure will be assessed as in the comparative approach, as well as plant production and the efficiency of nutrient use based on the use of ^{15}N-labeled plant material deposited at the soil surface, or as fertilizers inside the soil.

10.10.5 Diversity of Insects and Ecosystem Function

Hypothesis 9 The structural and chemical diversity of plants affects the richness and functional diversity of insects.

Rationale Herbivore insects are completely dependent on the quality and quantity of plant material. Evolution has led to a high degree of specialization in the use of particular resources by different group of insects.

Test 1. Correlative approach: the diversity of insects and plants might be recorded in a series of otherwise similar savannas (same climate and production), and the relationship may be established through statistical methods.

Test 2. Experimental approach: manipulation experiments with the removal of functional groups (insect populations are not altered by the disturbance itself).

Test 3. Natural experiments: analysis of insect populations in areas of savannas invaded by African grasses, compared to natural, nearby communities.

References

Anderson JM, Ingram JSI (eds) (1993) Tropical soil biology and fertility. A handbook of methods, 2nd edn. CAB International, Oxon

Baruch Z, Fernández D (1993) Water relations of native and introduced C_4 grasses in a neotropical savanna. Oecologia (Berl) 96:179-185

Belsky J (1990) Tree/grass ratios in East African savannas: a comparison of existing models. J. Biogeogr 17:483-489

Blanchart E, Lavelle P, Spain A (1990) Effects of biomass and size of *Millsonia anomala* (Oligochaeta: Acanthodrilidae) on particle aggregation in a tropical soil in the presence of *Panicum maximum*. Biol Fert Soils 10:113-120

Bormann FH, Likens GE (1979) Pattern and process in a forested ecosystem: disturbance, development and steady state based on the Hubbard Brook ecosystem study. Ecological Studies. Springer, Berlin Heidelberg New York

Braudeau E (1988) Méthode de caractérisation pédo-hydrique des sols basée sur l'analyse de la courbe de retrait. Cah ORSTOM Sér Pédol XXIV: 179-189

Chapin III FS (1991) Effects of multiple environmental stresses on nutrient availability and use. In: Mooney HA, Winner WE, Pell EWJ, (eds) Response of plants to multiple stresses. Academic Press, San Diego, pp 67-88

Frost P, Medina E, Menaut JC, Solbrig OT, Swift M and Walker BH (1986) Responses of savannas to stress and disturbance. Biol Int, Spec Issue 10. IUBS, Paris

Goldstein G, Sarmiento G (1987) Water relations of trees and grasses and their consequences for the structure of savanna vegetation. In: Walker BH (ed) Determinants of savannas. IRL Press, Oxford, pp 13-38

Hall DO, Long SP, Coombs J (eds) (1993) Photosynthesis and production in a changing environment: a field and laboratory manual. Chapman & Hall. London, New York

Hobbie SE, Jensen DB, Chapin III FS (1993) Resource supply and disturbance as controls over present and future plant diversity. In: Schulze E-D, Mooney HA (eds) Biodiversity and ecosystem function. Ecol Stud 99:345-408. Springer, Berlin Heidelberg New York

Hughes RF, Vitousek PM, Tunison JT (1991) Effects of invasion of fire-enhancing C_4 grasses on native shrubs in Hawaii Volcanoes National Park. Ecology 72:743-747

Huston M (1994) Biodiversity. Cambridge University Press. Cambridge

Lavelle P (1987) The importance of biological processes in productivity of soils in the humid tropics. In: Dickinson R E, Lovelock J (eds) Geophysiology of the Amazon. Wiley, New York, pp 175-214

Lavelle P, Martin A (1993) Small-scale and large-scale effects of endogeic earthworms on dynamics of organic matter of moist savanna soil. Soil Biol Biochem

Lawton JH, Brown VK (1993) Redundancy in ecosystems. In: Schulze E-D, Mooney HA (eds) Biodiversity and Ecosystem function. Ecological Studies 99. Springer, Berlin Heidelberg New York, pp 255-270

MacNaughton SAJ (1985) Ecology of a grazing ecosystem: the Serengetti. Ecol Monogr 55: 259-294

Medina E (1982) Physiological ecology of neotropical savanna plants. In: Huntley BJ, Walker BH (eds) Ecology of tropical savannas. Ecological Studies 42. Springer, Berlin Heidelberg New York, pp 308-335

Medina E, Silva JF (1990) Savannas of northern South America: a steady state regulated by water-fire interactions on a background of low nutrient availablity. J Biogeogr 17:403-413

Medina E, Bilbao B (1991) Signficance of nutrient relations and symbiosis for the competitive interaction between grasses and legumes in tropical savannas. In: Esser G, Overdieck D (eds) Modern ecology. Elsevier, Amsterdam, pp. 295-319

Menaut JC, Barbault R, Lavelle P, Lepage M (1985) African savannas: biological systems of humification and mineralization. In: Tothill JC, Mott J (eds) Ecology and management of world savannas. Aust Acad Sci, Canberra, pp 14-33

Muller RN, Bormann FH (1976) Role of *Erythronium americanum* Ker. in energy flow and nutrient dynamics of a northern hardwood forest ecosystem. Science 193:1126-1128

Pearcy R, Ehleringer J, Mooney HA, Rundel PW (eds)(1989) Plant physiological ecology: field methods and instrumentation. Chapman and Hall. London, New York

Sarmiento G (1983) Patterns of specific and phenological diversity in the grass community of the Venezuelan tropical savannas. J Biogeogr 10:373-391

Sarmiento G (1984) The ecology of neotropical savannas. Harvard Univ Press, Cambridge

Schulze E-D, Chapin FS III (1987) Plant specialization to environments of different resource availability. Ecol Stud 61:120-148

Spain A V, Lavelle P, Mariotti A (1992) Stimulation of plant growth by tropical earthworms Soil Biol Biochem 24:1629-1633

Stebbins GL (1952) Aridity as stimulus to plant evolution. Am Nat 86:33-44

Tilman D (1990) Constraints and trade-offs: toward a predictive theory of competition and succession. Oikos 58:3-15

Valentin C, Casenave A (1992) Infiltration into sealed soils as influenced by gravel cover. Soil Sci Soc Am J 56:1667-1673

Vitousek PM, Hooper DU (1993) Biological diversity and terrestrial ecosystem biogeochemistry. In: Schulze E-D, Mooney HA (eds) Biodiversity and Ecosystem function. Ecological Studies 99. Springer, Berlin Heidelberg New York, pp 3-12

Vitousek PM, Walker LR, Whiteaker LD, Mueller-Dombois D, Matson PA (1987) Biological invasion of *Myrica faya* alters ecosystem development in Hawaii. Science 238:802-804

Walker BH (1987) Determinant of savannas. IRL Press, Oxford

Walker BH (1992) Biodiversity and ecological redundancy. Conserv Biol 6:18-23

Walker BH, Menaut J-C (eds) Research procedure and experimental design for savanna ecology and management. RSSD, Australia. UNESCO-MAB

Walker BH, Noy-Meir E (1982) Aspects of stability and resilience of savanna ecosystems. In: Huntley BJ, Walker BH (eds) Ecology of tropical savannas. Ecological Studies 42. Springer, Berlin Heidleberg New York, pp 577-590

11 Biodiversity, Fire, and Herbivory in Tropical Savannas

11 Biodiversity, Fire, and Herbivory in Tropical Savannas

Bibiana Bilbao, Richard Braithwaite, Christiane Dall'Aglio, Adriana Moreira, Paulo E. Oliviera, José Felipe Ribeiro, and Philip Stott

11.1 Introduction

The main determinants of tropical savannas are plant available moisture (PAM) and plant available nutrients (PAN). Fire (F) and herbivory (H) are secondary determinants in a hierarchy of factors which produce the characteristics of any particular savanna (Solbrig 1991). While the biodiversity of a savanna area will be influenced by the biogeographic pool of species available to the region at any given time, the structure and productivity of the savanna are controlled by PAM and PAN, which in turn also influence biodiversity. Fire and herbivory then modify biodiversity directly through the mortality of individuals and indirectly through effects on the resources by individuals using different fuel types. Also biodiversity, through the range of food resource qualities and quantities, helps to determine the impact level of herbivory on the ecosystem.

11.2 Fire, Herbivory, and Biodiversity

Fire and herbivory are general pressures on savanna community organization, while other kinds of temporary pressures, such as episodic flooding and frost, are only locally important. The latter pressures may have an effect similar to that of fire and herbivory in the sense that they will change

Ecological Studies, Vol. 121
Solbrig, Medina and Silva (eds) Biodiversity and Savanna Ecosystem Processes

the heterogeneity of the community, altering species composition and structure. In the savannas of the Calabozo station (Venezuela), fire has had little impact on the herbaceous species richness. It was the evenness component of diversity alone which was affected by fire. Moreover, fire changed mainly the frequencies of the rare species, without altering significantly the dominance of the most common species in the area. Both fire history and species composition seem to affect fire impacts, but the changes in biodiversity were consistent. In savannas strongly dominated by *Hyparrhenia rufa*, the changes in evenness were greater than in savannas where the dominance was shared with native species of *Trachypogon* and *Axonopus*.

Biodiversity should be viewed as the independent variable affecting fire characteristics and impacts. Velocity, intensity, homogeneity, and other characteristics of the fire line are the result of at least three determinants, namely: weather conditions, topography, and fuel characteristics. Fuel density, height, and flammability will be determined by species composition and ultimately by biodiversity. Presence or absence of various functional groups also determines fire characteristics and their impacts. For example, *Pandanus* spp. acting as fuel ladders may take the fire into the canopy of certain Australian communities and the fire impact in such cases may be more severe. Differences in flammability between dried biomass of grass species also affects the homogeneity of the fire line, rendering a patchy fire impact in the area concerned.

We considered both aspects of fire/biodiversity relations and we maintain that fire impact on diversity may feed back through the characteristics of fire itself and then on the communities. Under fire conditions, survival of animals and maintenance of their biodiversity may depend on the diversity of shelters and possibilities of escape from fire. The presence of some functional groups in providing shelter for other species may be very important in this respect. For example, termite mounds and armadillo burrows are elements that increase shelter diversity and reduce fire impact on the fauna. Plant species may have similar effects and trunks or thick bark may provide refuges for elements of the fauna. These refuges and their protected populations may be important for the resource availability after fire and for the level of herbivory, and may have consequences for fuel accumulation and the subsequent fire regime.

The original patchiness of communities can result in a more heterogeneous fire behaviour and impact. Fire can increase resprouting and vegetative growth, resulting, perhaps, in a smaller intrapatch diversity and higher interpatch diversity. These changes will depend on the original patchiness and life-form composition of the area.

Another important topic is habitat diversity within an area and its importance for resilience under fire conditions. The mosaic of phyto-physiognomies, including gallery forests and less fire-prone physiognomies, may contribute to the survival of less fire-tolerant species of plants and animals. These species are able to recolonize other habitats after fire, increasing general diversity. At a regional scale, the relative species richness of the neighboring regions affects the changes in biodiversity under different fire regimes by increasing the chances of invasions. Species-poor rain forests around savanna areas in Australia, for example, may be the reason for the low rate of invasion in fire protected areas. In this sense, the size of fire disturbance and distance from more protected areas, besides dispersal rates, may also be important factors for recolonization and invasion after fire.

11.3 Fuel Mixture and Fire Diversity

The types and intensity of fires in savanna are mainly determined by the fuel mixture at the moment of the event.

Several factors contribute to characterize the fuel mixture:

1. Events occurring in the interval between fires: drought, above average rainfall, and herbivory.
2. Structure and composition of vegetation (biodiversity).
3. Spatial pattern of fuel distribution.
4. Fuel load.
5. Physiological development state of individuals.
6. Fuel quality and degree of flammability (amount of resins, lignin, cellulose, tannin, nutrient content, etc.)

In addition, site history will modulate directly and indirectly all aspects of the fuel mixture, such as live/dead relationship, moisture content, biomass load, relative abundance of different species, and biodiversity.

In terms of fuel mixture, we can define two important functional groups: the herbaceous and woody components. Density and spatial patterns of trees determine the spread of fire and its erratic performance (Stott 1986; Frost and Robertson 1987). Furthermore, the litter production contributes to the fuel load in the herbaceous stratum. Stott (1986) described in savannas of Thailand a typical litter burn, where the litter is the most important fuel. This type of fire presents a low fire-spreading rate and intensity.

Although the grasses dominate the herbaceous layer, there are other important elements such as legumes and other dicotyledonous species that contribute to the fuel load. In this sense we can characterize the biodiversity of fuel material (Bilbao 1995). Generally, the most intense and homogenéous fires are produced in communities in which there is a high biomass of grasses and a low fuel biodiversity (e.g., savannas dominated by *Hyparrhenia rufa* in Calabozo, Venezuela). In woodlands and shrublands where there are a lower grass biomass and a more uneven distribution of fuel, fires tend to be less intense and burn more patchily. Because of the greater flammability of the dead biomass, the dead/live ratio and the moisture content are a very significant factors determining fuel quality.

The residence time of fire depends on the spatial distribution of dead biomass. Therefore, the residence time will be lower in savannas dominated by grasses that have standing dead biomass exposed to the advance of fire. The woody component could slow down the fire, especially in areas where the trunks lie dead on the soil (Trollope 1978).

The plant architecture defines the vertical distribution of fuel. Taller grasses produce higher flames so that the probability of damage to the tree canopy will be higher (Frost and Robertson 1987). In this manner, the ladder effect could be seen in fires that occur in areas dominated by *Hyparrhenia rufa*, an African grass that reaches 2.5 m in Calabozo, Venezuela, while the native species, *Trachypogon plumosus*, does not grow over 1.5 m in the flowering period (Bilbao 1995).

Plant chemical composition is a very important characteristic that defines the fuel quality. Different plant compounds differ with respect to the degree of flammability. For example, lignin is very stable at high temperatures, losing only 30% of its weight at 400 °C (Philpot 1970), but cellulose and other carbohydrates have a much lower ignition point. The relative abundance of the most flammable compounds in the vegetation could lead to more intense fires (higher spreading rates). Several grasses and sedges protect their meristematic tissues from the effects of fire by means of "tunics," which are formed with the dead remainder of leaf sheaths or the external epidermis of rhizomes (Rachid-Edwards 1956). The low flammability of the tunics could be due to their chemical composition.

11.4 Hypothesis Formulation

In Chapter 10 it was argued that the development of hypotheses and experimental tests to measure the impact of biodiversity on the function of ecosystems is inherently difficult. Firstly, there are several methodological and conceptual problems associated with the estimation of biodiversity and, secondly it is extremely difficult to assess ecosystem function. Fortunately, fire is one of the few variables that can be readily manipulated and, as such, some experimental approaches regarding the effect of biodiversity on fire behaviour can be made.

Traditionally, in different studies concerning fire ecology, how fire intensity affects biodiversity was tested. However, we proposed a different point of view in which the species diversity plays an important role in ecosystem function.

11.5 Specific Hypotheses on Species Diversity and Fire Behaviour

Hypothesis 1 Biodiversity affects fire characteristics in tropical savannas.

Rationale Fuel density, structure, flammability, homogeneity, etc. will be determined by species composition and ultimately by biodiversity. Presence or absence of various functional groups also determines the fuel characteristics (e.g., relative dominance of grasses and trees, etc.). These fuel characteristics affect the fire behaviour (maximum temperature, rate of fire spread, flame height, etc.). Stott (1986) defined the vertical profiles of temperature burns in savannas of Thailand related to the presence of litter and the height and homogeneity of the grass layer.

Hypothesis 2 Biodiversity affects fire impacts on species composition in tropical savannas.

Rationale This hypothesis supports the idea that there is a feedback behaviour between fire and biodiversity in savanna ecosystems. In this manner, biodiversity determines the characteristics of the fire itself and then on communities. An example is in savannas dominated by *Hyparrhenia rufa* with a low biodiversity that produce fires of high intensity, allowing only the establishment of *H. rufa*, excluding the most sensitive species, and maintaining a low biodiversity. In contrast, savannas dominated by native South American species and with high biodiversity are characterized by fires of lower temperatures and fire-spreading rates with a low impact on species composition.

Hypothesis 3 Biodiversity affects fire impacts on availability of resources in tropical savannas.

Rationale In this case the impact is assessed on ecosystem function. Therefore fires in areas of low biodiversity would have a higher impact on the availability of resources than in more diverse communities.

Hypothesis 4 The patchiness of diversity of the community affects fire characteristics in tropical savannas.

Rationale Areas with a high patchiness show a high heterogeneity of the fuel material and produce fires of erratic performance and variable characteristics. The fire spread is not linear and presents several branches depending on fuel properties of the patches.

Hypothesis 5 The patchiness of diversity of the community affects fire impacts on patchiness of species composition in tropical savannas.

Rationale As in hypothesis 2, a feedback is contemplated in this proposition. Some authors consider that fire generates a higher patchiness in ecosystems. This will depend on the integrity, size, and biodiversity of the original patches and in the changes in composition due to processes other than herbivory and invasion of alien species.

Hypothesis 6 The patchiness of diversity of the community affects fire impacts on patchiness regarding the availability of resources in tropical savannas.

Rationale Sometimes it is difficult to separate direct and indirect fire effects on the availability of resources. For example, trees produce changes in the availability of soil nutrients and the accumulation of organic matter beneath their canopy. So, it is very important to define the limits of the patches and the different factors besides the fire that could contribute to the availability of resources.

Hypothesis 7 Habitat diversity affects the changes in biodiversity in areas with different fire regimes.

Rationale At a regional scale the relative species richness of neighboring regions affects the changes in biodiversity by increasing the chances of invasions and colonization of areas under different fire regimes. Moreover, the mosaic of phytophysiognomies, including gallery forests and less fire-prone physiognomies, may contribute to the survival of less fire-tolerant species of plant and animals.

References

Bilbao B (1995) Impacto del régimen de quemas en las características edáficas, producción de materia orgánica y biodiversidad de pastizales naturales en Calabozo, Venezuela. PhD Thesis, Instituto Venezolano de Investigaciones Cientificas, Caracas, Venezuela
Frost PGH, Robertson F (1987) The ecological effects of fire in savannas. In: Walker B. (ed) Determinants of tropical savannas. Monogr Issue 3:93-140, IUBS
Philpot CW (1970) Influence of mineral content on the pyrolysis of plant materials. For Sci 16:467-471
Rachid-Edwards M (1956) Alguns dispositivos para protecao de plantas contra a seca e o fogo. Bol Fac Fil USP 209, Bot 13:39-68
Solbrig OT (ed) (1991) Savanna modeling for global change. Biol Int Spec Issue 24:1-45, IUBS, Paris
Stott P (1986) The spatial pattern of dry season fires in the savanna forests of Thailand. J Biogeogr 13:345-358
Trollope WSW (1978) Fire behavior, a preliminary study. Proceedings of the Grasslands Society of Southern Africa 13:123-128

12 Savanna Biodiversity and Ecosystem Properties

12 Savanna Biodiversity and Ecosystem Properties

Steve Archer, Mike Coughenour, Christiane Dall'Aglio, G. Wilson Fernandez, John Hay, William Hoffmann, Carlos Klink, Juan F. Silva, and Otto T. Solbrig

12.1 Introduction

The overall question addressed here is the effect of different degrees of biodiversity on the function of savanna ecosystems. Function can be interpreted in two different ways. It can refer to the flow of energy and nutrients through an ecosystem or to the flow of species populations through time, i.e., the persistence of species populations and their properties, which we call the structure of the system. Here we discuss the effect of biodiversity on ecosystem function in this second sense. The role of biodiversity in the flow of energy and nutrients is addressed in Chapter 10.

When an ecosystem persists in time we say that it is stable. Stability may be the result of an environment that is invariant, or it may be due to the ability of species populations to persist (either by internal genetic adjustments, or by having broad tolerances) even when external conditions change. Stability has three components: resistance, resilience, and persistence. Each of these terms has a different meaning with respect to its interpretation of system dynamics following a perturbation.

Stability can be exhibited by different ecosystem properties. One measure of stability is floristic composition, including not only species combinations, but also diversity, i.e., species richness and evenness (for example as measured by the Shannon index or some other index, see Chap. 6). Demographic behavior, total biomass, cover, and similar system properties can also be used as measures of stability.

Ecological Studies, Vol. 121
Solbrig, Medina and Silva (eds) Biodiversity and Savanna Ecosystem Processes

While definitions of stability have previously carried implicit assumptions of an equilibrium or steady state as a reference point, more current definitions of stability recognize that a range or cloud of system states may be used for reference (Nicolis 1991, 1992). This range may contain regular cycles at different temporal scales, threshold responses, and apparently chaotic behaviors with underlying order (e.g., "strange attractors"). When cycling among system states is a characteristic system dynamic, it becomes essential to differentiate measures of short- and long-term stability, because while a short-term measure may indicate instability, a longer-term measure may indicate stability. While savannas may oscillate or fluctuate among a range of states, they still can be stable systems.

Interpretations of system stability responses depend upon the scale and level of organization of the stability measurements, as well as the expected system behavior at that scale or level of organization. Unfortunately, expected behaviors or states often erroneously assume that tree-dominated systems are the norm where mixtures of trees and other life-forms have persisted for millennia. Because interpretations of response measures are relative to some expected system behavior, they may be dependent upon the objectives of a study or upon the human uses of a system. It is important to recognize that the lack of a disturbance may constitute a perturbation if the system is normally subjected to disturbance. Therefore, the reference point or range of expected system dynamics needs to be explicitly stated and justified.

Measures of stability that are based upon floristic composition may provide results different from measures based upon functional group compositions due to functional redundancies among species within the functional groups. When species are more redundant in their functions, then it is less critical to maintain the same species, as long as all functions are preserved. Thus, the level of redundancy within groups must be known to interpret the significance of changes in floristic composition.

There are several important modifiers of disturbance responses. These modifiers must be taken into consideration to interpret responses to disturbances accurately. These are:

1. Time since disturbance.
2. Direct and indirect interactions among species.
3. Abiotic variables like soil depth, soil fertility, rainfall (PAM and PAN).
4. Other disturbances such as fire.

In particular, time-dependent variables like rainfall may confound responses to disturbance. When rainfall has changed over time since a disturbance, the effects of the change in rainfall must be disentangled from the effects of the disturbance in order to interpret the response.

It is important to recognize that the disturbance response depends upon the initial system state. In other words, system dynamics are sensitive to initial conditions. In the terms of sensitivity analysis, local sensitivity is likely to be different from global sensitivity. The shape of the global sensitivity response must be known in order to interpret system responses to disturbance. It is most likely that when there are a high number of species, there will be an increased probability that some of them might be adapted to the disturbance. If the system experiences a wide range of disturbances, then it may be more important that the system harbors even larger species pools. Density-dominance graphs can be used to determine if systems are under stress at the outset of the disturbance.

The history of disturbance has an important impact, through selective forces, upon the presence of species that are adapted to subsequent disturbances of the same kind. Indeed, savannas may be intrinsically stable relative to other systems, but only because they have evolved under disturbances like fire, herbivory and drought. Thus, continued persistence of savannas may necessitate continued disturbance to preserve stabilizing components and properties.

Modifiers of disturbance responses may be hierarchically ordered, with abiotic factors such as soils and climate constraining the effects of biotically mediated disturbances like herbivory or fire (Solbrig 1991). Other constraints on disturbance responses arise from evolutionary and biogeographic processes, inasmuch as these affect the pool of available species and their functional capabilities.

The invasion by alien species, particularly into South American savannas, is of great significance (see Chap. 5), as there are invasive challenges virtually everywhere that humans are present. Importantly, biotic diversity often decreases in systems that have been invaded, thus the implications for global biodiversity are great.

A number of important functional differences were pointed out between African invaders and South American native grasses (see Chap. 5). Thus, there are many possible secondary effects that may arise at the ecosystem level, via effects on herbivory, hydrology, decomposition, and nutrient cycling. Invaders may also initiate new successional processes due to their effects on abiotic and biotic processes. It is important, however, to demonstrate whether invasions are the result of differences in competitive abilities or if they are due to the primary disturbances of soil disruption and clearing of native species. Indeed, most invasions in South America have followed such primary disturbance of the soil and vegetation, and there is evidence that the invaders may not persist unless these anthropogenic disturbances continue. Furthermore, results from long-term fire exclosures in Venezuela demonstrate that African invaders are not as tolerant of fires as South American species, possibly due to the greater fuel accumulations they produce over long fire-free intervals. There is evidence from the

Serengetti in East Africa that native species are highly resistant to invasion on natural disturbances such as termite mounds, excavation patches of digging mammals, etc. While this may suggest that there may be differences among savannas in invasive resistance, there were invasions along roadcuts in the Serengetti.

12.2 Relationships Between Biodiversity and Stability: Some Hypotheses

Developing hypotheses and experimental tests to measure the impact of biodiversity on ecosystem stability is extremely difficult. In the first place, it requires a clear definition of biodiversity and an accurate way to measure it. As discussed by Luis Bulla in Chapter 6, there is no simple way of measuring diversity, all indices being fraught with methodological and conceptual problems. Furthermore, stability implies time, and as discussed above, long periods of time. We have no way of ascertaining the diversity of savannas in the immediate past, much less in the remote past. The time factor also makes any experimental approach to the testing of system stability extremely difficult.

The goal of the discussion group was to develop a set of hypotheses that would circumscribe the problem, and that would identify those properties of the system that should give instability. Experiments, observations, and computer simulations can then be designed to test the effects of these properties on the behavior of the system.

After discussing the comparative, experimental, and natural experiments approaches presented above (see Chap. 10), we add a fourth approach, computer simulation.

Computer simulation is no substitute for natural experiments and observations; but computer simulations can help to define parameters and test whether certain propositions are feasible. They are particularly useful when testing propositions that include random components and long time series.

12.3 Specific Hypotheses on Species Diversity and Ecosystem Stability

Global Hypothesis 1 Savannas are relatively stable systems, in terms of resistance and resilience. Savannas persist as savannas even though they experience an ongoing regime of disturbances from fire, drought, and herbivory.

Global Hypothesis 2 Savannas may, nevertheless, be pushed beyond the limits of their resilience and resistance domains into new configurations, by disturbances that are intense and uncharacteristic for the system.

Rationale These two general hypotheses state the perception based on the fossil record and their present broad distribution that savannas have persisted over a long time, at least since mid-Pliocene and probably before (van der Hammen 1983; Cole 1986). Yet, we know that some savannas can be transformed through human intervention into forests when fire is excluded (Chap. 2), or into grasslands through intensification of fire and other anthropogenic disturbances. These two hypotheses should then be seen as null hypotheses.

The term "stability " will be used to refer to both resistance and resilience components in the following hypotheses.

Hypothesis 1 Savannas are stable because a large fraction of nutrients, biomass, organisms, and meristems are located below ground, where they are protected from disturbance.

Rationale Savannas are exposed to frequent disturbances, especially fire. Since fire tends to be fast-moving and of short duration, it has a minimum effect on underground structures, and the large proportion of these may explain their stability.

Hypothesis 2 Savannas are stable because they are composed of a diverse array of species that have different functional properties. It would therefore be expected that there is a significant relationship between patterns of species richness and patterns of stability across sites within a savanna site as well as across savannas.

Rationale This hypothesis extends and generalizes hypothesis 5 of Chapter 10. We content that if a functional group is totally eliminated from the system, then the system will be more vulnerable to disturbance. So, for example, if trees are totally eliminated from a dry tree savanna, seedling establishment (that is concentrated under trees, Belsky 1990) will be reduced and the savanna will take longer to recover from the fire, or will not recover at all, changing into a grassland. Two additional hypotheses connected to Hypothesis 2 are:

Hypothesis 2a Savanna species are composed of disturbance-adapted species, along with less disturbance-adapted species, due to evolutionary histories of exposure to disturbance, and either cycles of disturbance/disturbance-free periods, or patchiness of disturbance and disturbance cycles in space. Thus, savannas will be less stable if disturbance-resistant (e.g., fire-resistant) or disturbance-sensitive species (e.g., shade-adapted species in a fire-prone system) are removed.

Hypothesis 2b Woody elements of savannas are more sensitive to disturbance than herbaceous components.

Hypothesis 3 The relationship between species or functional diversity in savannas is curvilinear, due to the bell-shaped curvilinear relationship between species numbers and PAM. Thus, savannas that experience intermediate levels of PAM may be expected to be most stable.

Rationale The text of the hypothesis is self-explanatory. It is difficult to test, however. It is amenable to computer simulation and we recommend that this avenue be explored.

Hypothesis 4 Relationships between stability and diversity are sensitive to spatial scale. Specifically, stability will be more likely as the scale (extent) of a system is enlarged . Conversely, systems that are highly fragmented will be less stable than connected systems. Stability is promoted by functional complementarity among different spatial components in the system, and interchanges of functionalities among different spatial components.

Rationale Savannas are very heterogeneous systems. They consist of areas of pure grassland, patches of trees and grasses, and small groves of trees. Furthermore, the density and importance of the tree and herbaceous component changes over the landscape in response to topographic, climatic, and disturbance features, and the history of the system. The smaller the time and/or spatial dimension of a savanna, the more likely it is that it will be modified as a result of a disturbance. For this reason, fragmentation should lead to loss of stability. Three corollaries of this hypothesis are:

Hypothesis 4a At larger spatial scales, there is a greater diversity of species and interactions among species, which increases the functional diversity of system as a whole (i.e., functions are integrated at larger scales to yield emergent, stabilizing properties at a higher level of system organization). Thus, gamma diversity will increase with scale, and this increase should be positively correlated with stability.

Hypothesis 4b Increases in topoedaphic diversity at patch, catena, landscape, and regional scales promote functional diversity among species, which then contributes to system level stability. Because topoedaphic diversity increases with scale, stability will likewise increase with scale.

Hypothesis 4c As the sample of savannas used to construct the relationship includes a larger number of sites, the relationship between diversity and stability will strengthen, due to stochasticity and weak relationships at smaller scales.

Hypothesis 5 Savannas may be pushed into new and stable configurations due to subsequent modifications of the environment by the biota, analogous to succession. The new system may have a similar diversity, but be composed of different species. The components of the new system may have different levels of resilience compared to the original components. The path of recovery back to the original composition may be different than the path subsequent to disturbance in terms of floristic succession (i.e., it is hysteric).

Rationale With time, savannas may change in species composition. This can result in environmental modifications that create new stable configurations, for example as a result of an increase/decrease in woody species that change the physicochemical composition of the soil (Chap. 3). Although the overall diversity may not change, other components of the system, such as resilience, may be different.

Hypothesis 6 The path of recovery back to the original composition of a savanna following a disturbance or a move to a new stable or quasi-stable state may be different than the path subsequent to a disturbance in terms of floristic succession (i.e., it is hysteric).

Rationale What this hypothesis states is that initial conditions determine the trajectory of the system.

Hypothesis 7 Savannas are stable because they are composed of species that are plastic in form and functionality.

Rationale We define plastic behavior as the ability of a species to modify its form/and or function in response to a disturbance, thereby reducing mortality. Plastic behavior in response to common disturbances, such as fire and drought, two common occurrences in savannas, is a major contributor to stability. A corollary is:

Hypothesis 7a Savanna species are more likely to exhibit plastic responses to disturbance or stress than in most other biomes.

Hypothesis 8 Savannas are stable because there are many functional groups, that when arranged along a continuum show a high degree of overlap among groups, and few spaces along the continuum that are devoid of function.

Rationale The more species overlap in their functional characteristics, the greater the probability that there will be a set of them capable of withstanding extreme disturbances, such as changes in climate, fire regime, or outside invasion. This hypothesis expresses in more general terms some of the statements made in Chapter 10.

References

Belsky J (1990) Tree/grass ratios in East African savannas: a comparison of existing models. J Biologeogr 17:483-489
Cole MM (1986) The savannas. Biogeography and geobotany. Academic Press, London, 438 pp
Nicolis G (1991) Non-linear Dynamics, Self-Organization and Biological Complexity. In Solbrig OT, Nicolis G (eds) 112 Perspective on Biological Complexity. IUBS, Paris, pp 7-50
Nicolis G (1992) Dynamical systems, biological complexity and global change. In: Solbrig OT, van Emdem HM, van Oordt PGWJ, Biodiversity and Global Change. IUBS Paris, pp 21-32
Solbrig OT (ed) (1991) Savanna modelling for global change. Biol Int Spec Issue 24:1-45, IUBS, Paris
Van der Hammen T (1983) The palaecology and palaeogrography of savannas. In: Bourlière F (ed) Tropical savannas. Elsevier, Amsterdam

13 Summary and Conclusions

13 Summary and Conclusions

Otto T. Solbrig

13.1 Introduction

We now try to answer the question we posed in the introduction: does biodiversity affect the functioning of tropical savannas. The answer is affirmative. Biodiversity affects the functioning of savannas through its effect on productivity (Chap. 2, 10), nutrient cycling (Chap. 3, 10), water economy (Chap. 4, 10), soil properties (Chap. 2, 3, 10) and possibly through its effect on long-term ecosystem resilience (Chapters 5, 9, 12).

Yet the near-term negative effects (costs) of converting savannas into pastures or agricultural fields are not very high and the benefits (income) obtained in increased productivity of those commodities that humans require or desire are usually sufficiently high that they outweigh the negative effects. The social and economic forces operating today are such that the conversion of tropical savannas into pastures or agricultural fields is likely to continue. The degree of change and the forces accounting for these changes vary from continent to continent (Young and Solbrig 1993). In India, savanna transformation is already very advanced; in Africa, the encroachment on savanna lands comes primarily from a population displaced by a more technological, capital-intensive, labor-saving agriculture in the better lands, and from overall demographic pressures in a continent with the highest growth rate in the world; in the Americas the llanos and especially the cerrados are increasingly being transformed for soybean cultivation and cattle raising. Because of low precipitation and very poor soils, agriculture or intensive cattle raising in Australian Savannas is generally noneconomic and the transformation of the savannas for agriculture or cattle raising is only possible in parts of Queensland.

Ecological Studies, Vol. 121
Solbrig, Medina and Silva (eds) Biodiversity and Savanna Ecosystem Processes

The transformation and concomitant loss of biodiversity and its implications for ecosystem function all over the world are very worrisome for what they imply in terms of ecosystem function loss in the future. Since tropical savannas occupy about 40% of the surface of the geographical tropics, the loss or drastic curtailment of their biodiversity is especially disquieting. Brazilian savannas are a good illustration of likely future scenarios (Klink et al. 1993; Cunha 1994; Mueller 1995)

13.2 The Transformation of the Brazilian Savannas: an Evaluation

The cerrado region in Brazil occupies an area of 1.55 million km², about 20% of the surface of Brazil. It extends in a broad arch from near the Atlantic coast south of the mouth of the Amazon river in the northeast of Brazil to the border of the Pantanal in the cental-east of the country (Fig. 13.1). These are the richest savannas of the world in terms of species, with about 10 000 species of trees and shrubs, and several thousand species of herbs (Ratter 1986) The cerrado extends from near sealevel to over 1500 m, and the topography is that of a plain with a low undulations and elevated plateaus. The soils are oxisols, with a low content of clay, fine texture, very low in nutrients especially phosphorus and nitrogen, very low pH, low cation exchange capacity, and often a high aluminum content. The climate is typical of tropical savannas with a 5 to 7 month wet period followed by 7 to 5 months of drought. Overall, yearly precipitation varies from 800 to over 1500 mm. The wet period is likely to be interrupted by a short period of drought that can last for 1 to 3 weeks, especially in the more southern parts of the area.

Cultivation of these soils cannot be done without massive capital investments, calculated in 1987 to be about 800$ per ha (Goedert 1990). The land is first cleared of the woody vegetation with the help of bulldozers and is then leveled. It is then heavily limed to increased the pH and neutralize the negative effects of aluminum toxicity. It is then given a fertilizationed with phosphate rock. The lime and phosphate treatment has to be repeated every 10 to 15 years. Finally, a yearly application of nitrogen fertilizer is required. In some areas low in boron, addition of this element is also required.

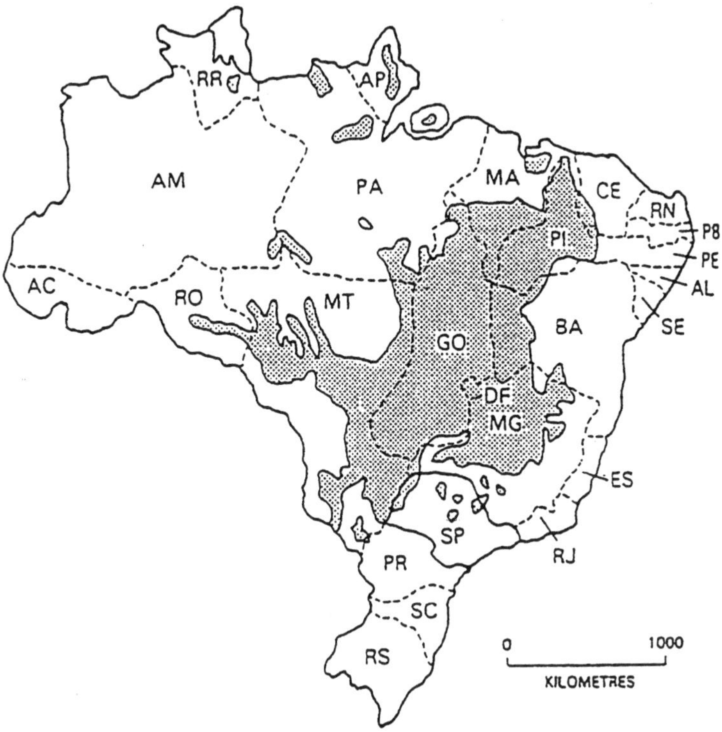

Fig. 13.1. Map of Brazil with the region of the cerrados indicated by *stripples*. The map shops also the political boundaries of the states. The *abbreviations* correspond to the names of the states of Brazil

Table 13.1 Cultivated surface and production of principal crops in the cerrado (1990)

Crop	Area (ha)	Production (Tm)
Soybeans	3 365 000	5 048 000
Maize	1 745 000	3 403 000
Rice	1 054 000	980 000
Beans	345 000	244 000
Coffee	171 000	246 000
Manihot	101 000	1 267 000

Source: IBGE (Instituto Brasileiro de Geografiae Estadistica), 1990. Produção Estatistica Municipal. Brasilia

Under these conditions productivity can be high, especially with soybeans (Table 13.1) Yet the total production of all commodities in the 6 790 000 ha taken over for agriculture (some of which was originally gallery forest land) was 11 188 000 Tm (Mueller 1995). If we assume a natural above-ground productivity of the herbaceous stratum of 1000 gm^{-2} $year^{-1}$, which is very low for a tropical savanna (Sarmiento 1984; Long et al. 1992), the production is at least six times less than what would have been produced naturally. This calculation does not take into account that the actual productivity of the agricultural fields is two to five times higher since only the products harvested by humans are counted in the 6 790 000 Tm. However the natural productivity is also an underestimate, since it does not take into account the woody component, and uses a very low productivity value for the grasses. At most, the agricultural fields harvest only half as much light energy as would have been the case had the area not been modified, energy that would have maintained a larger biomass of consumers and decomposers, and would have contributed more carbon for soil organic matter formation.

A different situation holds for the area planted with forage plants. Although there are no data on actual primary productivity, secondary productivity in these pastures is double that in unimproved rangelands (24.5 animals km^{-2} vs. 10.7 animals km^{-2}). Some of the increased productivity is the result of the species composition of the planted pasture, but there also might be an increase in actual productivity as the result of the better nutrient status of the soil (Fisher et al. 1994).

Soil erosion in cultivated fields, especially soybean fields, is a serious problem in Brazil. It is so serious that it is leading to the abandonment of cultivation in nonflat areas, especially Minas Gerais. The major problem is water erosion resulting from the almost daily torrential summer rains. Common conservation techniques used in temperate regions, such as contour plowing, are ineffective in tropical situations and may be counterproductive, since the furrows act like small dams, and when they brake, more water rushes down the hill with devastating effects. Minimum tilling and deep plowing to increase drainage are resisted by Brazilian farmers because of increased costs and the special machinery that is required (Cunha 1994). Soil erosion also exposes plintite layers that on exposure become hardened making the fields impossible to cultivate.

To contrast these negative effects agricultural research institutions in the area, most notably the Brazilian agricultural research system (EMBRAPA, Empresa Brasileira de Pesquisas Agronomicas) and the International Tropical Research Center in Cali, Colombia (CIAT, Centro Internacional de Agricultura Tropical), have embarked in research programs to develop more sustainable techniques, with some success.

However, the objective of sustainable development is not only to develop technologies that reduce soil loss but to "attend to the needs of the poor of the world" who according to the World Commission on Environment and Development (WCED 1987) "should be given the first priority." According to the same source, sustainable development should also recognize the existence within a given technology of environmental limitations. In this context, the situation in Brazil is not sustainable, since agricultural development, especially soy bean cultivation, is an export-driven development, highly capitalized, which extracts energy and materials from the cerrado ecosystem to export mainly to Europe where most of the soybeans are transformed into animal feed, while many from the low-income population of Brazil suffer malnutrition. Yet the production costs of Brazil's soybean agriculture are such that they are beyond the buying power of Brazil's poor.

13.3 The Future of Savanna Biodiversity

The economic and social forces that are driving the transformation of the natural savanna landscape - indeed of all natural landscapes - are such that they cannot be easily stopped as long as the stated goals of most people of the world and their political representatives are to increase consumption under the guise of a supposedly "higher standard of living." The prevailing economic model, and the global integration of the economy into one world economy directed at satisfying the consumption goals of the industrialized countries and local country elites, are nonsustainable. Ever-increasing consumption in the developed world rather than rising populations in the developing world (although they also contribute their share) is responsible for the despoiling of the natural landscapes of the world. Unless a political will develops to change the present value system and to replace it by a more equitable, efficient, and environmentally sustainable system (Young and Solbrig 1992, 1993), which does not appear to be very likely, the reduction in savanna biodiversity, and especially in the delivery of ecological services by tropical savannas, will continue with the ultimate result of lowering the carrying capacity for humans of these systems.

References

Cunha A S (1994) Uma availacao da sustentabilidade da agricultura nos cerrados. IPEA, Brasilia 200 pp

Fisher MJ, Rao IM, Ayarza MA, Lascano CE, Sanz JI, Thomas RJ, Vera RR (1994) Carbon storage by introduced deep-rooted grasses in the South American savannas. Nature 371:236-238

Goedert W (1990) Estrategias de manejo das savannas. In: Sarmiento G (ed) Las sabanas americanas. CIAT, Merida, Venezuela, pp 191-218

Klink CA, Moreira AG, Solbrig OT (1993) Ecological impact of agricultural development in the Brazilian cerrados. In: Young MD, Solbrig OT (eds) The world's savannas. Economic driving forces, ecological constraints, and policy options for sustainable land use. Parthenon, Paris, pp. 259-283

Long SP, Jones MB, Roberts MJ (1992) Primary productivity of grass ecosystems of the tropics and subtropics. Chapman and Hall, London

Mueller CC (1995) A sustentabilidade da expansao agricola nos cerrados. Instituto Sociedade, Populacao e Natureza, Documento de Trabalho 36:1-21

Ratter JA (1986) Notas sobre a vegetacao de fazenda Agua Limpa (Brasilia, DF) con una chave para os generos lenhosos de dicotyledoneas do Cerrado. Editoria Univ Brasilia, Brasilia

Sarmiento G (1984) The ecology of neotropical Savannas. Harvard University Press, Cambridge, 235 pp

Young MD, Solbrig OT (1992) Savanna management for ecological sustainability, economic profit and social equity. Mab Digest 13:1-47

Young MD, Solbrig OT (eds) (1993) The world's savannas. Economic driving forces, ecological constraints, and policy options for Sustainable Land Use. Parthenon, Paris 350 pp

World Commission on Environment and Development (WCED) (1987) Our common future. Oxford Univ Press, Oxford

Subject Index

Subject Index

Ecological Studies

Volumes published since 1990

Volume 80
Plant Biology of the Basin and Range (1990)
B. Osmond, G.M. Hidy, and L. Pitelka (Eds.)

Volume 81
Nitrogen in Terrestrial Ecosystems: Questions of Productivity, Vegetational Changes, and Ecosystem Stability (1991)
C.O. Tamm

Volume 82
Quantitative Methods in Landscape Ecology: The Analysis and Interpretation of Landscape Heterogeneity (1990)
M.G. Turner and R.H. Gardner (Eds.)

Volume 83
The Rivers of Florida (1990)
R.J. Livingston (Ed.)

Volume 84
Fire in the Tropical Biota: Ecosystem Processes and Global Challenges (1990)
J.G. Goldammer (Ed.)

Volume 85
The Mosaic-Cycle Concept of Ecosystems (1991)
H. Remmert (Ed.)

Volume 86
Ecological Heterogeneity (1991)
J. Kolasa and S.T.A. Pickett (Eds.)

Volume 87
Horses and Grasses: The Nutritional Ecology of Equids and Their Impact on the Camargue (1992)
P. Duncan

Volume 88
Pinnipeds and El Niño: Responses to Environmental Stress (1992)
F. Trillmich and K.A. Ono (Eds.)

Volume 89
Plantago: A Multidisciplinary Study (1992)
P.J.C. Kuiper and M. Bos (Eds.)

Volume 90
Biogeochemistry of a Subalpine Ecosystem: Loch Vale Watershed (1992)
J. Baron (Ed.)

Volume 91
Atmospheric Deposition and Forest Nutrient Cycling (1992)
D.W. Johnson and S.E. Lindberg (Eds.)

Volume 92
Landscape Boundaries: Consequences for Biotic Diversity and Ecological Flows (1992)
A.J. Hansen and F. di Castri (Eds.)

Volume 93
Fire in South African Mountain Fynbos: Ecosystem, Community, and Species Response at Swartboskloof (1992)
B.W. van Wilgen et al. (Eds.)

Volume 94
The Ecology of Aquatic Hyphomycetes (1992)
F. Bärlocher (Ed.)

Volume 95
Palms in Forest Ecosystems of Amazonia (1992)
F. Kahn and J.-J. DeGranville

Volume 96
Ecology and Decline of Red Spruce, in the Eastern United States (1992)
C. Eagar and M.B. Adams (Eds.)

Volume 97
The Response of Western Forests to Air Pollution (1992)
R.K. Olson, D. Binkley, and M. Böhm (Eds.)

Volume 98
Plankton Regulation Dynamics (1993)
N. Walz (Ed.)

Volume 99
Biodiversity and Ecosystem Function (1993)
E.-D. Schulze and H.A. Mooney (Eds.)

Volume 100
Ecophysiology of Photosynthesis (1994)
E.-D. Schulze and M.M. Caldwell (Eds.)

Ecological Studies

Volumes published since 1990

Volume 101
Effects of Land Use Change on Atmospheric CO_2 Concentrations: South and South East Asia as a Case Study (1993)
V.H. Dale (Ed.)

Volume 102
Coral Reef Ecology (1993)
Y.I. Sorokin

Volume 103
Rocky Shores: Exploitation in Chile and South Africa (1993)
W.R. Siegfried (Ed.)

Volume 104
Long-Term Experiments With Acid Rain in Norwegian Forest Ecosystems (1993)
G. Abrahamsen et al. (Eds.)

Volume 105
Microbial Ecology of Lake Plußsee (1993)
J. Overbeck and R.J. Chrost (Eds.)

Volume 106
Minimum Animal Populations (1994)
H. Remmert (Ed.)

Volume 107
The Role of Fire in Mediterranean-Type Ecosystems (1994)
J.M. Moreno and W.C. Oechel

Volume 108
Ecology and Biogeography of Mediterranean Ecosystems in Chile, California and Australia (1994)
M.T.K. Arroyo, P.H. Zedler, and M.D. Fox (Eds.)

Volume 109
Mediterranean-Type Ecosystems. The Function of Biodiversity (1995)
G.W. Davis and D.M. Richardson (Eds.)

Volume 110
Tropical Montane Cloud Forests (1995)
L.S. Hamilton, J.O. Juvik, and F.N. Scatena (Eds.)

Volume 111
Peatland Forestry. Ecology and Principles (1995)
E. Paavilainen and J. Päivänen

Volume 112
Tropical Forests: Management and Ecology (1995)
A.E. Lugo and C. Lowe (Eds.)

Volume 113
Arctic and Alpine Biodiversity. Patterns, Causes and Ecosystem Consequences (1995)
F.S. Chapin III and C. Körner (Eds.)

Volume 114
Crassulacean Acid Metabolism. Biochemistry, Ecophysiology and Evolution (1995)
K. Winter and J.A.C. Smith (Eds.)

Volume 115
Islands. Biological Diversity and Ecosystem Function (1995)
P.M. Vitousek, L.L. Loope, and H. Adsersen (Eds.)

Volume 116
High Latitude Rainforests and Associated Ecosystems of the West Coast of the Americas: Climate, Hydrology, Ecology and Conservation (1995)
R.G. Lawford, P. Alaback, and E.R. Fuentes (Eds.)

Volume 117
Anticipated Effects of a Changing Global Environment on Mediterranean-Type Ecosystems (1995)
J. Moreno and W.C. Oechel (Eds.)

Volume 118
Impact of Air Pollutants on Southern Pine Forests (1995)
S. Fox and R.A. Mickler (Eds.)

Volume 199
Freshwater Ecosystems of Alaska (1996)
A.M. Milner and M.W. Oswood (Eds.)

Volume 120
Landscape Function and Disturbance in Arctic Tundra (1996)
J.F. Reynolds and J.D. Tenhunen (Eds.)

Volume 121
Biodiversity and Savanna Ecosystem Processes. A Global Perspective (1996)
O.T. Solbrig, E. Medina, and J.F. Silva (Eds.)

MIX
Papier aus verantwortungsvollen Quellen
Paper from responsible sources
FSC® C105338

If you have any concerns about our products,
you can contact us on
ProductSafety@springernature.com

In case Publisher is established outside the EU,
the EU authorized representative is:
Springer Nature Customer Service Center GmbH
Europaplatz 3, 69115 Heidelberg, Germany

Printed by Libri Plureos GmbH
in Hamburg, Germany